The Evolving Universe

The Evolving Universe

The Evolving Universe, Relativity, Redshift and Life From Space

GA Mohr, Richard Sinclair, Edwin Fear

Copyright © 2014 by GA Mohr, Richard Sinclair, Edwin Fear.

Library of Congress Control Number: 2014908340
ISBN: Hardcover 978-1-4990-0208-9
Softcover 978-1-4990-0207-2
eBook 978-1-4990-0206-5

All rights reserved. No part of this book may be reproduced or transmitted in any form or by any means, electronic or mechanical, including photocopying, recording, or by any information storage and retrieval system, without permission in writing from the copyright owner.

Any people depicted in stock imagery provided by Thinkstock are models, and such images are being used for illustrative purposes only. Certain stock imagery © Thinkstock.

Rev. date: 05/12/2014

To order additional copies of this book, contact:
Xlibris LLC
1-800-455-039
www.Xlibris.com.au
Orders@Xlibris.com.au
620769

Contents

Preface 9

1. The Atom 13
2. The Earth 22
3. The Solar System 34
4. The Universe 43
5. The Interstellar Medium 54
6. The Theory of Relativity.. 64
7. The Steady-State Theory of the Universe................ 80
8. The Big Bang Theory 87
9. Is the Universe Really Expanding? 94
10. Is the Universe Infinite? 104
11. The Cosmic Microwave Background 109
12. Scientific Theories of the Universe 115
13. Religious Beliefs 124
14. Is There Life Out There? 130
15. The Evolving Universe 142
16. Conclusions 155

Appendix A: Earth's Evolutionary Timeline................ 165
Appendix B: Gravitational Deflection of Light............ 168
Appendix C: The Large Curvature Correction
   In FEM, by G. A. Mohr 174
Appendix D: The Ridiculous Theory of Relativity 181
Appendix E: Many Remaining Questions 187

Glossary... 191
References 199

Also by G. A. Mohr and Edwin Fear

*World Religions: The History,
Psychology, Issues, and Truth*

Also by G. A. Mohr

A Microcomputer Introduction to the Finite Element Method

Finite Elements for Solids, Fluids, and Optimization

*Curing Cancer and Heart Disease,
Proven Ways to Combat Aging,
Atherosclerosis and Cancer*

*The Pretentious Persuaders
A Brief History and Science of Mass Persuasion*

*The Variant Virus
Introducing Secret Agent Simon Sinclair*

*The Doomsday Calculation,
The End of the Human Race*

*The War of the Sexes
Women Are Getting on Top*

*Heart Disease, Cancer, and Aging,
Proven Neutraceutical and Lifestyle Solutions*

The History and Psychology of Human Conflict

Preface

I have long had an interest in science in general, no doubt encouraged from the outset by my science background—my mother, having been a chemist, and my father, one of the team that split the atom in Cambridge in 1931. Myself, in fact, I spent much of my adult life researching the finite element method, and a slightly relevant appendix on that subject is appended to this book.

In my penultimate Leaving Certificate year at Melbourne Grammar School, I chose Fred Hoyle's book *The Black Cloud* as my prize for excellence in Mathematics. That book, of course, was fiction, but I have retained a passing interest in his steady-state theory of the universe over the years whilst viewing, as do many other people, the Big Bang theory with some scepticism.

At the present time, however, there is growing dissent with the Big Bang theory, in part because some of its adherents also propose that the universe might end with another 'singularity', a Big Crunch.

The 'everything from nothing' idea of the Big Bang and the 'everything to nothing' idea of the Big Crunch are unacceptable proposals to many people. For example, in speaking to a lady friend in Oxford a couple of months ago, she said something along the lines that it is difficult to imagine such an event as the Big Bang resulting in a world so full of life as that we live in.

Perhaps the foundation stone of the Big Bang theory is the belief that the universe is expanding. This belief is based on Hubble's law, which attributes 'redshift' of radiation from distant stars to longer (redder) wavelengths to the Doppler effect, which assumes that these stars are speeding away from us. As the

Preface

redshift is greater for more distant stars, Hubble's law makes their recession velocity proportional to their distance from us.

This is an absurd idea, and indeed, Hubble originally favoured an alternative idea—the 'tired light' theory—to explain redshift. One key point made in this book, however, is that the most logical explanation for redshift is cosmic dust, of which there is a great deal in the 'interstellar medium', and particularly in our galaxy.

Another key foundation of the Big Bang theory is the relatively uniform cosmic microwave background (CMB) radiation discovered in 1964. This is attributed to an incredibly short period of 'inflation' supposed to have spread matter from an infinitesimal 'singularity' all over the universe in a relatively uniform manner.

Again, however, cosmic dust is the more likely explanation, perhaps in conjunction with a couple of other mechanisms discussed in Chapter 11.

For such reasons, this book cautiously proposes the 'evolving universe' theory of the universe. This proposes that from as far back in time as our observations allow us to see, the universe has been slowly approaching some sort of equilibrium or steady state. Here it should be noted that how far back we can see relates to the speed of light and the distance of the farthest galaxies that we can see.

The measure of the universe's evolution is taken to be the rate of star formation in the universe, which is believed to have peaked around eleven billion years ago and to be a fraction of that peak level now. Before that time there may have been a long period of growth towards the peak rate of star formation, though this poses the question: Why was there such a period and what caused it?

With any theory of the universe, the questions of, "Why is there anything at all?" or "Where did it come from?" cannot be answered, and the main point of the present theory is not to deal with these impossible questions but merely to say to what situation the universe is progressing.

Thus, comparable to the evolution of life on Earth seeming to have reached some sort of a peak, so too the star formation rates in the universe have been gradually decreasing, hopefully to some stable equilibrium value.

Preface

Such a model, crude as it is, does stand in stark contrast to proposals of a Big Crunch that will collapse the universe, or cyclic expansion and contraction of the universe over many tens of billions of years.

Perhaps the most important contributions of this book, however, are proposed alternative, more rational explanations for such phenomena as redshift and the cosmic microwave background radiation (CMB).

Discussion of the possibility that life on Earth evolved from bacteria that arrived on meteors from space is also a key point of interest, one for which there is now much supportive evidence.

The book also briefly discusses the theory of relativity, this having played an important role in modern cosmology. As with the Big Bang theory, there is growing discontent with the special theory of relativity, particularly such predictions as clocks in two frames of reference both being able to be considered to be moving slower than the other.

Buenker's alternative Lorentz transformation (ALT) is thus briefly discussed, as this preserves 'simultaneity of events', avoiding one of the major paradoxes involved in the special theory of relativity.

Finally, the authors are grateful to Annie Maynard and the other staff of Xlibris for their excellent and efficient work in publishing this book.

GAM 2014

Chapter 1
The Atom

> The atoms struggle and move in the void because of dissimilarities between them and other differences; and as they move they collide and become entangled in such a way as to cling in close contact to one another.
>
> Democritus (attributed).

The periodic table

Modern science has developed an increasingly sophisticated picture of the matter of which the universe is composed. The Greek philosopher Democritus (460-370 BC) was perhaps first to envisage matter as consisting of very small atoms, but it was not until the eighteenth and nineteenth centuries that the science of chemistry developed to the point at which a large number of elements had been identified.

In 1869, Dmitri Mendeleev published the first widely recognized periodic table, which illustrated periodic trends in the properties of the then well-known elements. He also predicted the existence of a number of elements needed to fill gaps in his table, and these were subsequently discovered (Medeiros 1971).

Chapter 1

Mendeleev's table listed the elements in order of increasing atomic weight, but the modern periodic table lists them in order of increasing atomic number (the number of protons in the nucleus), the rows of the table being called periods and some of the columns being called groups.

Since that time many more elements have been discovered and today the periodic table lists 118 elements, 98 of which exist naturally, the others being radioactive elements with short half-lives synthesized in laboratories.

The Bohr-Rutherford model of the atom

Neils Bohr briefly worked with Ernest Rutherford in Manchester in 1912 before returning to Copenhagen to marry. In 1913, he modified the Rutherford nuclear model of the atom by assigning electrons orbits of fixed size and energy, the basis for his model being the quantized energy emissions from black bodies discovered by Planck in 1900, which were the foundation for the development of quantum mechanics.

In 1919, Rutherford produced artificial transmutation of atoms for the first time using alpha particles (i.e., helium nuclei) from radium to transform nitrogen atoms into oxygen atoms (Gaines 1970):

$14N7 + 4He2$ transforms to $1H1 + 17O8$

In 1931, Cockroft and Walton were first to split the atom by using artificially accelerated particles (protons) for the first time to split a Lithium isotope into two alpha particles:

$7Li3 + 1H1$ transforms to $4He2 + 4He2$

This development led, of course, to the development of the atomic bomb and nuclear power.

By then, therefore, the proton-neutron-electron model of the atom was fully established.

The Atom

New atomic particles

The electron has mass 0.511MeV and charge 1.6×10^{-19} Coulomb, whereas proton mass is 936MeV (1.67262×10^{-27} kg) with opposite charge.

The electron-neutrino was proposed in 1930, and the neutron was discovered in 1932. Artificially produced free neutrons can break down into a proton, an electron, and an antineutrino (the antimatter counterpart of the neutrino with no charge and little mass).

Antimatter particles have the same mass and electric charge of ordinary matter such as electrons and protons but have opposite charge and magnetic properties. Antimatter includes the antiparticles positrons, which were detected in cosmic rays in 1932, and antiprotons and antineutrons, which were produced in particle accelerators. Antimatter exists only briefly, being quickly annihilated by contact with normal matter, when large amounts of energy are released.

Models of the atom have greatly increased in complexity in the last several decades. The Standard Model of fundamental particles brackets electrons and neutrinos as being *leptons,* and includes two types of *quarks,* each having a three member family as shown in Table 1.1. (Oerter 2006):

Table 1.1. The lepton and quark families.

	Family		
Leptons	1	2	3
Neutrinos	electron neutrino	mu neutrino	tau neutrino
Electron and relatives	electron	muon	tau
Quarks	up	charm	top
	down	strange	bottom

Only the particles of the first family are stable, the others being short-lived by-products of high-energy collision. Muons and taus, for example, have the same charge and spin as the electron and mass 106 MeV and 1,777 MeV respectively, having lifetimes of 2.2 microseconds and 10^{-13} seconds. Both have associated neutrinos.

The existence of the tau neutrino was confirmed experimentally in July 2000.

Quarks are fundamental particles that interact via the *strong force* to bind the components of the nucleus. Only the up and down quarks are needed to make protons and neutrons, the others occurring only in heavier, unstable particles.

Quarks have mass and spin but no apparent structure, so they cannot be resolved into something smaller. Quarks always seem to occur in combination with other quarks or antiquarks.

Bosons are the third family of fundamental subatomic particles, the others being leptons and quarks. There are three types of bosons: photons, gluons, and weakons or weak bosons. Photons are light particles that transmit electromagnetic forces, gluons transmit forces between quarks, and weakons change one type of particle into another.

Gluons are massless particles with one unit of intrinsic spin that carry the strong force which is about 100 times greater than the electromagnetic force (emf) within its short range of about 10^{-15}m. In strong interactions quarks exchange gluons, which act as bundles of energy that bind protons and neutrons.

The Higgs boson, sometimes called "the God particle" because some believe it played a key part in the Big Bang, was detected in July 2012 at CERN. The Higgs boson interacts with fundamental particles to give them their mass, and the Higgs mechanism explains why carriers of the short range *weak nuclear force* are heavy while the photons which carry the long range emf are massless.

In radioactive decay, the strength of the weak force is about 1/100,000 that of the emf, but intrinsically, the weak force has as much effect as the emf, and according to 'electroweak theory', the two are different forms of a single electroweak force.

One predicted type of boson, the massless graviton which transmits gravity with unlimited range, remains undetected as yet.

Nucleosynthesis

Nucleosynthesis is the synthesis on a cosmic scale of chemical elements from as few as one or two simple types of atomic nuclei, a process involving large-scale nuclear reactions such as those that occur in the sun and other stars, 98% of the mass of which are made up of hydrogen and helium. Thus heavier chemical elements are formed from the most basic ones such as hydrogen and helium by the addition of extra protons and neutrons to their nuclei.

Thus by repeated nuclear fusion, four hydrogen nuclei combine to form a helium nucleus. In turn, carbon and oxygen can be formed by 3 and 4 helium nuclei respectively. In this way, the most abundant elements up to iron (atomic number 26) can be formed (Hoyle 1983). Indeed, Rutherford's transmutation of nitrogen into oxygen, the equation for which was given earlier, could be taken as an example of such processes.

Elements heavier than iron can then be formed by capture of a neutron, subsequent beta decay converting the extra neutron into a proton, and emitting an electron and antineutrino.

Table 1.2. Nucleosynthesis reactions in a star about twenty-five times the mass of the sun.

Stage	Temp. deg. C x 10^6	Density kg/cc	Duration of stage
H —> He	40	0.005	10^7 years
He —> C, O	200	0.7	10^6 years
C —> Ne	600	200	600 years
Ne —> Mg	1,200	500	1 year
C, O —> Si, S	1,500	10,000	6 months
Si —> Fe	2,700	30,000	1 day
Core collapse	5,400	3×10^8	0.25 secs
Core bounce	23,000	3×10^{11}	0.001 secs
Explosive	> 1,000	varies	10 secs

Chapter 1

Table 1.2 shows the different stages of nucleosynthesis estimated for a star of 25 solar masses. The successive nature of such processes is the reason why the relative abundance of elements decreases as their atomic number increases. Thus the solar system contains about a million times more carbon, nitrogen, and oxygen than much heavier elements such as gold and platinum.

In addition, stable isotopes with even numbers of protons and neutrons occur more often than those with odd ones, this being called the 'odd-even effect'. Thus, of almost three hundred stable nuclides (nuclei with a specific number of protons and neutrons) known, just five have odd numbers of both protons and neutrons, and more than half have even atomic and mass numbers (respectively the number of protons, and protons plus neutrons in the nucleus).

In addition, nuclides that have equal and even numbers of neutrons and protons, for example the 'alpha-particle' nuclides such as carbon-12, magnesium-24 and argon-36, have considerable stability and comparatively high abundance.

Finally, elements with atomic numbers 2, 10, 18, 36, 54, and 86 have complete electron shells and are thus more stable, the noble gases helium, neon, and argon being examples of this.

The supply of nuclear energy in larger stars is proportional to their energy, but their mass-luminosity relation involves a high power of their mass so that larger stars have shorter lifetimes. Indeed, the brightest stars in existence at present may only be a few million years old.

It is only in larger stars that the heavier elements can be produced, and it is believed that there is much exchange of matter between stars and the interstellar medium and that nucleosynthesis in heavier stars produced some of the heavy elements found today.

The explosion of supernovas, stars reaching the end of their evolution, is probably the main process by which heavy elements produced inside stars are returned to the interstellar medium, from which stars are formed.

Cosmic rays

Another means of propagation of matter in space is cosmic rays. Cosmic rays are high-speed nuclei and electrons that travel throughout the galaxy at about 87% of the speed of light. Their composition is 87% protons, 12% alpha particles, 0.25% light elements (lithium, beryllium, and boron), the remainder being electrons and nuclei of heavier elements.

A small proportion come from the sun, but most are believed to come from the galaxy's disc and supernovas' explosions and the residues left by these. Cosmic rays are relatively rich in heavier elements compared to normal matter, suggesting that their earlier stages of acceleration do indeed occur in regions rich in heavy elements.

Another source of cosmic rays may be X-ray binaries such as Cygnus X-3. In these systems, a normal star loses mass to a companion neutron star or black hole, both forms of collapsed star.

Lower-energy cosmic rays are very sensitive to interplanetary gravitational fields so that most of those detected in the earth's upper atmosphere have very high energy and have an isotropic spatial distribution.

Cosmic-ray particles that arrive at the top of the Earth's atmosphere are termed primaries, and their collisions with atmospheric nuclei give rise to secondaries.

From the 1930s to the 1950s, cosmic rays were the only source of high-energy particles used in the study of particle physics because they impart more energy than most particle accelerators.

Quantum radiation

The Bohr model of the atom explains the three types of spectra:

[1] An *emission line* is produced when an electron jumps from an orbit with energy E_2 to an orbit with a lower energy E_1, emitting a photon with energy $E = hf = E_2 - E_1$ where h is Planck's constant and f is the frequency of emission.

Chapter 1

[2] An *absorption line* occurs when an atom with an electron in orbit with energy E_1 absorbs a photon with energy $E = E_2 - E_1$, jumping the electron to higher energy level E_2. The photon is later re-emitted but in a random direction so that the part of the spectrum associated with energy E is darkened.

[3] A *thermal spectrum* is produced when atoms are closely packed together, energy levels of atoms being distorted by electrons in neighbouring atoms. This widens the normally sharp spectral lines, and when the density is high enough, a continuous spectrum with the full 'rainbow' of colours is produced.

Conclusion

The electron was discovered in 1897, the nucleus in 1911, protons and neutrons in the 1930s, and quarks in the 1970s. Since the 1960s, a number of types of meson have been discovered, including the pi meson, or pion, which plays a role in the strong interactions between protons and neutrons in nuclei. Mesons are composed of an even number of quarks and antiquarks and are unstable.

More than two hundred subatomic particles have been detected thus far, most appearing to have a corresponding antiparticle. Thus the Standard Model of atomic physics is close to finality and does much to explain the complex internal mechanisms at work in nuclei.

Another important advance in atomic physics has been an understanding of the nucleosynthesis that occurs in stars to form heavier elements.

The Big Bang model of the history of the universe assumes very rapid creation of a quark-gluon plasma. Then, as the temperature of the universe cooled, quarks were amalgamated by massless gluons to form protons and neutrons (three quarks each) and then simple nuclei.

With further cooling, the simplest atoms hydrogen and helium were able to form and, as cooling continued, hot clouds of these were able to cluster to form the first suns (Singh 2005).

The Atom

The recently detected Higgs boson, which imparts mass to fundamental particles such as quarks, completes the standard atomic model and thus our understanding of the Big Bang model. As will be discussed in later chapters, however, creation of a seemingly infinite universe from a 'singularity' involving infinite density and temperature should be impossible and more rational and probable models than the Big Bang have been proposed.

Chapter 2

The Earth

> We live between two worlds; we soar in the atmosphere;
> we creep upon the soil; we have the aspirations of
> creators and the propensities of quadrupeds.
> There can be but one explanation of this fact.
> We are passing from the animal into a higher form,
> and the drama of this planet is in its second act.
>
> W. Winwood Reade, *The Martyrdom of Man*.

Earth's formation

Our solar system probably began about 4.6 billion years ago as a coalescence of dust and gases from an exploding star. The gravitational pull being strongest at the centre, thermonuclear reactions began in the compressed and superheated gases. This was the beginning of our sun where these reactions converted hydrogen into helium and released intense radiation.

Around the sun, dust particles combined, eventually forming rocky fragments or planetesimals (very early meteorites). These collided, sometimes disintegrating, and sometimes combining into larger masses. As the largest of these masses grew, their increasing gravity attracted more planetesimals, and eventually, the first planets were formed, leaving impact craters as evidence of their violent formation (Brown and Morgan 1989).

The Earth

Asteroid data, along with discoveries of 'gas giants' orbiting close to alien suns as a result of migration, have led to the alternative 'Nice model' of planet formation. This suggests that gas giants formed close to our sun and migrated to their current positions through gravitational interactions, scattering large populations of asteroids and comets and creating a Late Heavy Bombardment of the inner planets and moons about 3.9 million years ago.

At first, the Earth was relatively cool, but continuing meteorite impacts formed a sea of molten rock covering a cool interior. Slowly the interior was heated by the decay of radioactive elements and iron from impacting meteorites sank to form a molten metallic core.

On the surface water, complex carbon compounds, and even amino acids were deposited by meteorites. Radioactive minerals near the surface decayed into stable elements and water vapour, carbon dioxide (CO_2), nitrogen and other gases rose to form thick clouds which gravity prevented from escaping into space.

Within a billion years of its formation, the planet cooled sufficiently for water vapour to fall as rain and iron-rich sedimentary rocks were formed at the bottom of the seas that formed. On the newly solidified continents, lakes formed and rivers carried minerals to the sea. Supplied by terrestrial weathering and undersea volcanic eruptions, the oceans acquired an increasingly complex array of chemicals. Lightning in the upper atmosphere resulted in 'nitrogen rains', from which the first simple organic chemicals were formed. Indeed, as proof of this process, organic chemicals have been formed in laboratories by exposing an approximation of the earth's early atmosphere and ocean to electrical charges.

The beginning of life

Water molecules were among the original compounds in the solar system and were dispersed in meteorites to all the planets, but only Earth has liquid water because of its size and distance from the sun, which keeps the medium atmospheric temperature at a cool 60 °F.

Chapter 2

At the beginning of life, the earth's atmosphere is thought to have been similar to those still found on Mars and Venus, high in CO_2 but with little free oxygen. This lack of oxygen was important at this stage because the earliest organic cells had not developed defences against oxidation.

It has been claimed that the ancestral gene of all life on Earth comes from bacteria that grow in boiling water with traces of sulphuric acid, and similar organisms can still be found in volcanic regions such as Mount St Helens and Yellowstone (Brown and Morgan 1989).

It is more likely that the first forms of life developed about 3.8 billion years ago and were prokaryotes developed in warm ponds or deep-sea hydrothermal vents, perhaps as both blue-green algae and bacteria (BBC 2014, Smithsonian Institute 2014).

Others believe that the life arrived on Earth in bacterial form from the interstellar medium or aboard meteorites, and this issue is discussed further in Chapter 14 (Hoyle 2013, Jayawardhana 2011)

Bacteria-like organisms preserved in rock dating back some 3.5 billion years have been found near the North Pole, and a few fossils have been found in Australia dating back 2.5 billion years.

As far back as 4 billion years ago, cyanobacteria (blue-green algae) produced the complex organic molecule chlorophyll, which uses sunlight to fuel the unique process of photosynthesis which converts abundant carbon dioxide and hydrogen into carbohydrate and releases oxygen. As a result, oxygen levels in the atmosphere reached the present level of about 21% about a billion years ago.

Stromatolites are seabed collections of cyanobacteria that were once the planet's dominant life form and major source of oxygen. During the day, they use sunlight to synthesize glucose and calcium carbonate from water and carbon dioxide, at night secreting the calcium carbonate to form cementitious layers that eventually harden into rock.

Bacterial and viral evolution

From the primitive early forms of bacteria that were the first life forms, countless types of bacteria have evolved, for example the methanogen family of bacteria that operate in the stomachs of ruminant animals to convert hay, carbon dioxide, and hydrogen into protein and methane.

Now every handful of soil contains millions of bacteria, and our bodies are filled with a multitude of different types of bacteria.

A modern example of bacterial evolution came with the discovery of antibiotics, the widespread use of which led to the development of antibiotic-resistant bacterial forms such as MRSA (methicillin-resistant staph aureus). The appearance of bacteria resistant to the most powerful antibiotics such as vancomycin is now of great concern.

New viruses have also evolved, for example several strains of avian flu, the most recent being H9N2, which was first discovered in Hong Kong in 2009; and there were unconfirmed reports of a man having contracted this virus in Hong Kong in 2013 (ABC News24, 1PM, Dec. 31, 2013).

In the late nineteenth century, Antoine Bechamp named *microzymas* after the Greek words, meaning 'small' and 'ferment'. He claimed that these were the essential unit of life and studied them in various diseases, including TB. He found that microzymas could ferment, except at high temperatures.

Bechamp tried to kill microzymas but found them indestructible. They were present in the amoeba, the smallest form of animal life, and in bacteria, the smallest form of plant life. Under certain conditions, they transformed into bacteria, first enlarging into round forms, two or more of these, then coupling to form rods.

In the 1930s, Wilhelm Reich named the microbes he found in cancer tumours *bions*. He found that they survived at 1500 °C, concluding that they were a preliminary form of life in transition between inorganic and organic.

He named the bacteria that grew in sterile cancer tissue *T-bacilli*, eventually finding these cancer microbes in cancer tumours and also in the blood and excreta of healthy people.

The blood of cancer patients, however, produced T-bacilli easily whereas normal blood produced them slowly. Nevertheless, Reich had shown that viral cancer may not be exogenous but arise from within.

Several other researchers have also found cancer microbes (Cantwell 1990).

In 1987, Dr Tania Korsak concluded from her studies of the pleomorphic tubercle that rod-like forms in culture break down into smaller coccoid forms. She concluded that the coccoidal microbial forms found in AIDS and cancer are not staphylococci but a type of virus.

It can be concluded therefore that (Mohr 2013a):

(a) bacteria and viruses are not distinct and, indeed, one can transform into the other.
(b) under adverse conditions pleomorphic microbes can change form and initiate cancer.

Thus, having been the planet's first life forms, bacteria play an important role in its ecosystem; and, of course, 'good bacteria' play an important role in our digestive system, whereas 'bad bacteria' can cause serious diseases, including cancer.

Evolution of plants

All plants and animals derived from bacteria-like organisms that evolved more than 3 billion years ago, plants evolving from algae and evidence of the first land plants dates back about 500 million years.

More than 2 million species of plants and animals have been identified, and it is believed that many more are yet to be discovered. These have great heterogeneity of size and shape, from bacteria less than a thousandth of a millimetre in diameter to trees rising 100 metres into the air.

The habitats of these creatures also vary greatly, from bacteria living in near-boiling hot springs to fungi and algae living in Antarctica, and from wormlike creatures thousands of feet below the surface of the ocean to spiders and plants at more than 20,000 feet above sea level on Mount Everest (*Encyclopaedia Britannica* CD 1999).

The evolutionary history of plants is recorded in fossils found in lowland or marine sediments. Such fossils show the external form of the plants, and sometimes cellular structures. Other microfossils are pollen and spores.

The fossil record indicates increasing rates of evolution and increasing diversity and complexity that began with invasion of the land and colonization of the continents.

The first land plants were probably the bryophytes, comprising the nonvascular mosses, liverworts, and homworts, and most fossils of these are similar to living forms.

Next came herbaceous weeds, which covered bare earth, enriching the environment ready for the development of larger herbs, perennial shrubs, trees and vines.

The first fossil evidence of land plants dates back to circa 470 million years ago, and the first megafossil evidence of plants dates back to about 420 million years ago, these plants being only a few centimetres in height.

The first primitive vascular plants appeared about 400 million years ago and colonized moist habitats. These had axes with terminal sporangia and stomata and were thus photosynthetic. By about 380 million years ago, the first plants with leaves appeared and soon thereafter shrub-like plants developed.

These were limited in size by the limited diameter that their stems and roots could achieve. The ancestors of the seed plants overcame this restriction by developing bark so that by about 340 billion years ago, forests comparable to those we see today began to develop.

The largest extant seed plants are the gymnosperms including conifers and cycads, but many other forms of these existed during the Mesozoic era 245 to 66.4 million years ago. From some of these early forms of flowering plants (angiosperms) developed, and today, there are about 250,000 varieties, making up 90% of vascular plants (*Encyclopaedia Britannica* CD 1999).

Indeed, evidence of Earth's abundant plant life is found in its spectrum, as observed by orbiting spacecraft, which shows a 'bump' and a steep 'red-edge' due to chlorophyll (Jawawardhana 2011).

Chapter 2

Evolution of animals

Many forms of phyla appeared in the Cambrian period (544-505 million years ago), including the first vertebrates, but most of these became extinct. Fossils from the beginning of this period include calcium carbonate shells; it is thought because by this time, oxygen levels had become sufficient to support complex animals and the collagen needed to form hard body structures.

In the early Cambrian period, there was no life on land except cyanobacteria in moist sediments, and only a few basic species lived in the open sea.

On the margins of the oceans, however, the aquatic ecosystem included the relatively large carnivore Anomalocaris and early arthropods which fed upon marine deposits.

In the Ordovician period (505-438 million years ago), the marine fauna included bryozoans, brachiopods, corals, cephalopods, and crinoids, but environmental changes at the end of this period eliminated many marine species.

During the Silurian period (438-408 million years ago), the predatory cephalopods became abundant, and many new suspension feeders and the earliest craniates appeared.

The first amphibians and insects appeared during the Devonian period (405-345 million years ago), the first reptiles appearing about 300 million years ago. Mammals including dinosaurs appeared during the Triassic period (230-190 million years ago), dinosaurs becoming extinct during the early Jurassic period (190-135 million years ago), some believe because of impact by a cluster of meteorites or because of a brief surge in volcanic activity (Brown and Morgan 1989).

Primates appeared circa 130 million years ago, grazing and carnivorous mammals about 35 million years ago and the first humans some 1.6 million years ago (*Encarta 99*).

Chimpanzees, with whom we share 98% of the same genes, are quite sociable animals and live in groups of twenty to sixty, forming into subgroups of adults (male and female), all-male groups, and groups of mothers and offspring. African gorillas also live in bisexual groups of between two to thirty but which do not comprise smaller subgroups.

The best-known studies of chimpanzees were conducted by Jane Goodall and associates in the Gombe National Park on the edge of Lake Tanganyika in Tanzania (van-Lawick-Goodall 1971).

Ultimately, Goodall was quite disillusioned to find that tribes of chimps were led by an alpha male and would occasionally have small wars with neighbouring tribes, these resuming at intervals over periods of many years. She concluded that they were all too much like humans!

Comparisons of blood proteins and the DNA of the African great apes with that of humans indicate that the line leading to modern people did not split off from that of chimpanzees and gorillas until comparatively late in evolution, perhaps 6 million to 8 million years ago.

Evolution of man

Charles Darwin's 1859 book, *On the Origin of Species by Means of Natural Selection*, established the theory of evolution firmly whilst the botanical studies of Alfred Russel Wallace helped broaden and strengthen that theory. The sciences of genetics and molecular biology have since provided strong confirmation of it.

An overwhelming body of scientific evidence continues to accumulate about the evolution of modern humans from chimpanzees, with whom we share circa 98% of our DNA. A key step in this evolution was the appearance of hominids with bipedal locomotion between 5 and 6 million years ago in Africa, remains of the first *Ramapithecus* species of *Hominidae* having been found in Kenya dating back to the middle of the Miocene epoch, that is, about 19 million years ago (Weiss and Mann 1978).

According to fossil records, about 5 million years ago, the first of seven species of australopithecines appeared in Africa, there being two types, the robust and the gracile australopithecines. These species were from 1.2 to 1.4 metres tall and weighed from 30 to 45 kg. The robusts, as their name suggests, were more solidly built but became extinct about a million years ago.

The more fleet-footed graciles survived, however, evolving from the *Australopithecus afarensis* form to the species *Australopithecus africanus* by about 2.5 million years ago (Smith and Davies 2008).

Chapter 2

The earliest evidence of stone tools comes from sites in Africa dated to about 2.5 million years ago. These tools have not been found in association with a particular hominine species.

About 2 million years ago Australopithecus africanus evolved into the first *Homo* species, *Homo habilis*, the forerunner of modern man. *Homo habilis* evolved into *Homo ergaster* about 1.5 million years ago in Africa and spread into Asia, where it evolved into *Homo erectus*, a species which survived until about 250 thousand years ago (*Encarta* 2009).

Later *H. erectus* skulls possess brain sizes in the range of 1100 to 1300 cc (67.1 to 79.3 cu in), within the size variation of *Homo sapiens*.

A number of archaeological sites dating from the time of *Homo erectus* reveal a greater sophistication in toolmaking than was found at earlier sites. Evidence found at the cave site of Peking man in northern China suggests that *H. erectus* used fire.

The remains of the foundations of an oval structure built by a *Homo erectus* group were found at the Terra-Amata site in France, and within this structure, there was a fireplace (Weiss and Mann 1978).

Homo ergaster then spread from Africa into Europe, evolving into *Homo heidelbergensis*, so named because the first remains of this species were discovered in Heidelberg, Germany, in 1903. This species appeared between 0.6 and 1.3 million years ago and survived until 200 to 250 thousand years ago.

The *Homo* species spread widely, and by 350,000 years ago planned hunting, fire making, wearing of clothes, and probably burial rituals, were well established.

It seems likely that *Homo heidelbergensis* then evolved into *Homo sapiens* Neanderthalis between 200,000 and 300,000 years ago. The Neanderthals had similar DNA to modern man and lived only in family groups, the men being hunter-gatherers to feed the family.

The Neanderthals left cave paintings, which were an important evolutionary advance. These often depicted a simple activity, perhaps a precursor to the highly pictorial hieroglyphic script of the ancient Egyptians (Egerton Eastwick 1896).

Meanwhile, in Africa, *Homo ergaster* evolved into *Homo sapiens sapiens* at around the same time, spreading to Europe and interbreeding with the Neanderthals so that circa, 4% of the DNA of non-African modern humans comes from them.

The Neanderthals had slightly larger brain size than *Homo sapiens* sapiens, but disappeared about 30,000 years ago, in part as a result of interbreeding.

Fragments of another subspecies of *Homo sapiens*, the Denisovans, dating back 40,000 years, were recently discovered in Siberia along with Neanderthal remains. Study of the nuclear genome of this species suggested that it came from the same origins as the Neanderthals. The Denisovans ranged from Siberia to Southeast Asia, and up to 6% of their DNA is found in Melanesians, Australian Aborigines, and the Mananwa, a Negrito people of the Philippines.

Comparison of the Denisovan and Neanderthal genomes showed that there was considerable interbreeding between the two species, the Denisovan DNA being 17% Neanderthal.

Some scientists believe in the 'replacement model', which holds that Neanderthals were replaced by migrating *Homo sapiens sapiens*. As noted above, however, the evidence now supports the 'assimilation model' in which there was a significant amount of interbreeding.

The 'assimilation' or 'multiregional evolution model' proposes that modern humans evolved more or less simultaneously in the major regions of the world; for example, modern Chinese are thought to be evolved from archaic Chinese humans.

The present author believes this is true in this instance at least, and that modern Chinese people evolved from the *Homo erectus* species that evolved from the spread of the *Homo ergaster* species from Africa to Asia.

Like chimpanzees, *Homo sapiens sapiens* formed tribes, and there is evidence of religion, recorded events, and art dating from 30,000 to 40,000 years ago, implying the advanced language and ethics required for the ordering of social groups.

An example of relatively recent evolution, the body shapes of people in central Africa contrast greatly with those of Eskimos, the latter being shorter and carrying much more body fat to cope with cold climate (Weiss and Mann 1978).

Chapter 2

Conclusions

Earth itself evolved in a fashion, its moderate temperature preserving an abundance of water from which sprang the first life forms. Millions of years ago, these made their way onto land, eventually spreading widely and evolving into an enormous variety of forms, as summarized in Appendix A.

Indeed, evolution is still continuing, an example being butterflies inhabiting industrial areas becoming black in order to become more tolerant of pollution and less conspicuous to predators. Another example is the evolution of insect strains resistant to insecticides, house flies having become highly resistant to DDT in many parts of the world.

Indeed, man's interference in the global ecosystem is likely to have devastating effects ultimately, some experts predicting mass extinction of more than half the world's plant and animal species within the next 200 years.

In addition, new viruses, such as Golden Staph, Hep C, VRE, Aids, Ebola, Marburg, and avian flu may eventually spin out of control and further decimate both animal and human populations (Mohr 2012).

A further consequence of man's excessive population is that of 'reverse evolution', evidence of which includes reductions in IQ in the last century, observed in the UK and US, dramatic increases in the incidence of most types of cancer, often as the result of defective genes (Weinberg 1999), and male sperm counts decreasing by 50% in Copenhagen in the last fifty years and by 40% in the last twenty years only in Paris (Mohr 2012).

Indeed, Mohr (2012) calculates that, thanks to man's gradual degradation of the planet in various ways, our population will peak in about 200 years and then decline greatly until it reaches a level about half that at present.

Indeed, half the present population is all that is realistically obtainable if everybody old enough is to live in a nice house with most of the mod cons, have a PC, a mobile phone, a 'tablet', and a car, et cetera.

The Earth

Furthermore, given the degree to which we have already exhausted many of the earth's finite resources and polluted, warmed, and desertified the planet, it may be that even with only half the present population in a few hundred years' time, we will likely still experience the same problems of poverty and inequality that have plagued us for millennia.

Chapter 3
The Solar System

> *The day of the sun is like the day of a king. It is a promenade in the morning, a sitting on the throne at noon, a pageant in the evening.*
>
> Wallace Stevens, *Souvenirs and Prophecies*.

The sun

As noted at the beginning of Chapter 2, our solar system began as a cloud of super hot gas and dust some 4.6 billion years ago. At first much of this condensed to form the sun, which contains 99.9% of the mass of the solar system, in like fashion to the nucleus of the gold atom containing 99.95% of its mass (Oerter 2006).

In stars as massive as the sun, or greater, the sequence of nuclear fusion reactions required to produce many of the heavy elements can occur. Some stars steadily lose mass, in part by expelling material into the interstellar medium, but supernova explosions are the main process by which heavy elements are returned to space.

The first nuclear reaction in stars is conversion of hydrogen into helium, and a thin shell of helium separates the region where hydrogen has not yet been converted into helium from that where helium has been fused into heavier elements.

The Solar System

Thus the central region of a highly evolved star contains heavier elements such as iron and nickel with layers of successively lighter elements surrounding it, and the outermost layer contains mostly hydrogen or hydrogen and helium.

The sun is entirely gaseous, mostly hydrogen with a significant proportion of helium, along with small amounts of heavier elements in gaseous form. It has a very dense core where thermonuclear fusion produces helium from hydrogen, along with electromagnetic radiation, much of this the visible light that falls on Earth.

The sun's diameter is almost 1.4 million km, which is approximately 110 times that of the earth. It has a mass of approximately 2×10^{30} kg, about 330,000 times the mass of the Earth. The average density of the sun, however, is only about 1/3 that of the Earth, though the sun's core has a much greater density than Earth's core. Indeed, the great majority of the sun has a density much less than that of Earth's atmosphere (Jones 2010).

Because the 99.9% of the mass of the solar system is found in the sun, hydrogen and helium account for 98% of the mass of the solar system. Then thanks to the successive nature of nucleosynthesis, the solar system contains about a million times more carbon, nitrogen, and oxygen than the much heavier elements such as gold. This is because conversion of lower masses to higher masses is usually far from complete, only a limited proportion of neutron captures by nuclei being accompanied by conversion of the additional neutron into a proton.

Thermonuclear fusion of hydrogen

The density of the core of the sun, which fuses hydrogen as its energy source, is approximately 150 gm/cc, 150 times that of water and 10 times that of lead. This great density forces particles closer together, making reaction more likely.

The sun's core temperature is 15 million °K; this high temperature is giving particles high velocity, which makes them more likely to combine.

Chapter 3

Thermonuclear fusion of hydrogen requires three steps. The first is beta decay when two hydrogen nuclei (or protons) collide and one proton decays into a neutron, a positron (anti-electron) and an electron neutrino, the neutron combining with the other proton to form a deuterium or 'heavy hydrogen' nucleus.

In the second step, another proton combines with the deuterium nucleus to form an unstable helium-3 nucleus, resulting in gamma ray emission.

The first two steps having occurred numerous times, the third step is combination, with the emission of two protons, of two He3 nuclei, to form a stable He4 nucleus. The net result of the three steps can be written as

4 protons = He4 + 2 gamma ray photons

And the overall reaction is a good source of energy because the mass of four H nuclei (protons) is greater than the mass of one He4 nuclei.

The planets

The size and distance from the sun of the eight planets and two of the five dwarf planets of the solar system are shown in Table 3.1. (Lofts, Preuss, and Gilbert 1991, Quercus 2013).

Thousands of asteroids less than 950 km in diameter also orbit the sun, most of them between Mars and Jupiter and most of them only several kilometres in diameter.

Earth rotates once every 24 hours, Mars once every 24.5 hours, and Jupiter once every 10 hours. Venus rotates in the opposite direction to all the other planets and takes 243 days to do so. The rotation periods (length of day) and time periods for orbiting the sun (length of year) for the planets are shown in Table 3.2. (Lofts, Preuss, and Gilbert 1991, Sparrow 2013).

The Solar System

Table 3.1. Planets of the solar system.

Planet	Diameter at equator (Earth = 1 unit)	Average distance from sun (Earth = 1 unit)	Satellites
Mercury	0.38	0.37	0
Venus	0.95	0.71	0
Earth	1.00	1.00	1
Mars	0.53	1.50	2
Jupiter	11.2	5.18	16
Saturn	9.5	9.47	18+
Uranus	3.7	19.0	15
Neptune	3.5	30.0	5
Pluto	0.2	39.4	3
Eris	0.18	67.6	1+

Table 3.2. Rotation and orbit periods for the planets.

Planet	Rotation period	Sun orbit period
Mercury	59 days	88 days
Venus	243 days	225 days
Earth	1 day	1 year
Mars	24.5 hours	687 days
Jupiter	10 hours	11.9 years
Saturn	10.7 hours	29.5 years
Uranus	16 hours	84 years
Neptune	16 hours	165 years
Pluto	6.4 days	248 years
Eris	> 0.33 days	557 years

Most of the planets have moons or satellites moving around them. Phobos, the smaller of Mars' two moons, is only 9 km high and 11 km across. The largest moon in the solar system is Jupiter's Ganymede, which is 1.5 times the size of Earth's moon.

Chapter 3

Many comets also orbit the sun at speeds between 1000 and 2 million km/hr, their periods of orbit varying from several years to millions of years. The solar system also contains interstellar gas and dust and meteorites; the largest meteorite to fall on earth having weighed 59 tonnes.

Mercury has no atmosphere but that of Venus's is thick clouds of mostly carbon dioxide. The Martian atmosphere is thin and mostly carbon dioxide. The atmospheres of Jupiter, Saturn, and Uranus are mostly hydrogen and helium, that of Jupiter's being turbulent, and that of Uranus's including clouds of methane. The atmosphere of Neptune is mostly hydrogen with clouds of methane while Pluto has no atmosphere.

The masses and densities of the sun, the eight planets, and two of the dwarf planets are shown in Table 3.3 (Quercus 2013).

Table 3.3. Masses and densities of the planets.

Planet	Mass (Earths)	Density (Water = 1)
The Sun	333,000	1.41
Mercury	0.055	5.43
Venus	0.816	5.24
Earth	5.96 X 10^{21} tonnes	5.52
Mars	0.108	3.93
Jupiter	318.3	1.33
Saturn	95.1	0.69
Uranus	14.5	1.32
Neptune	17.1	1.64
Pluto	0.002	2.06
Eris	0.003	2.35

The sun's diameter is 1,390,000 km, compared to 12,756 for the Earth, and thence it's far greater mass.

The relatively small masses of Pluto and Eris show why they are classified as dwarf planets.

The temperature and gravity of the planets vary considerably, as shown in Table 3.4 (Lofts, Preuss, and Gilbert, 1991).

The Solar System

Table 3.4. Temperature and gravity of the planets.

Planet	Surface temperature °C	Surface gravity X that of Earth
Mercury	-180 to 420	0.38
Venus	Average 480	0.91
Earth	Average 17	1
Mars	-120 to 20	0.38
Jupiter	Average—140	2.6
Saturn	Average—170	1.2
Uranus	Average—210	0.93
Neptune	Average—220	1.2
Pluto	Average—230	0.05

Our solar system is now regarded as having only eight planets and five dwarf planets, including Pluto.

The solar system contains many other objects of significance, including an asteroid belt between the orbits of Mars and Jupiter, and the Kuiper belt, a ring of debris including many comets outside the orbit of Neptune.

Newly found planetary objects

The inner asteroid belt, which includes the mini-planet Ceres, separates Mercury, Venus, Earth, and Mars from the four middle giants: Jupiter, Saturn, Uranus, and Neptune.

Between Saturn and Uranus is the planetoid Chiron.

The outermost solar system begins beyond the orbit of Neptune with the Kuiper asteroid belt, which contains a number of small planets and extends 30 to 50 AU from the sun. With a diameter of 2,300 km, the dwarf planet Pluto is one of the largest Kuiper belt objects, along with Eris, which is of similar size.

In the vicinity of Pluto are many smaller objects, including Quaoar, Varuna, Orcus, Ixion, Haumea, and Makemake. Dozens of other objects are hundreds of kilometres across, and more, are discovered each year, three recently discovered objects being proposed as possible dwarf planets.

Chapter 3

Rodney Gomes, an astronomer at the National Observatory of Brazil in Rio de Janeiro, analysed the orbits of 92 Kuiper belt objects and found that half a dozen, including the remote body known as Sedna, have orbits that appear distorted, perhaps by an as yet undetected faraway planet that is large enough to exert gravitational effects upon Kuiper belt objects. Gomes calculated that the mystery planet might be about four times bigger than Earth and 1,500 times farther away from the sun than Earth.

Another possibility, of course, might be the presence of a concentrated mass of dark matter somewhere near the Kuiper belt, dark matter having been proposed as the additional matter needed to explain observed gravitational effects in the solar system and beyond.

Outside the Kuiper belt, a void is followed by a relatively unexplored area of space leading to the Oort cloud, a hypothetical huge collection of comets far beyond the orbit of Pluto.

The Milky Way

Our sun is one of some 100,000 million stars in the Milky Way, which also contains vast amounts of interstellar gas and dust from which new stars are continually formed.

There are two basic types of galaxies, elliptic and spiral, the Milky Way being the latter. Other types are irregular galaxies, lenticular galaxies (with a central region), radio galaxies, and Seyfert galaxies, which have weak spiral arms and a small bright nucleus (Jones 1987).

Looking along the plane of the Milky Way, one sees countless stars combining to form a band of misty light.

The Milky Way has three main regions:

[1] The *central bulge* contains old, relatively cool stars, and very little interstellar gas and dust.
[2] The *disc* contains the four spiral arms, which consist of younger, much hotter stars and much more interstellar material.

The Solar System

[3] The *halo* surrounds the central region of the galaxy and contains very old stars concentrated in clusters which move around the galactic centre in highly elongated orbits.

The sun is 30,000 light years or about two-thirds of the way out from the centre near the second Sagittarius constellation. The nearest star to the sun is Proxima Centauri, 4.3 light years away, whereas the sun is just 8 light minutes away from Earth.

Measuring distances in space

Distances of celestial objects can be estimated from parallax observations. Observing an object from two widely separated points, the angle between the two lines of sight is twice the parallax. Solar parallax, for example, is a maximum of 8.794 seconds of arc when observers are diametrically opposed.

For the most distant objects, parallax is calculated from the difference in a star's position when seen from Earth at points six months apart in its stellar orbit. Alpha Centauri, the brightest star in the constellation Centaurus, has a stellar parallax of 0.76 seconds of arc.

Is the solar system expanding?

If the solar system is expanding, the rate of expansion is bound to be very slow because the gravitational binding in the solar system cluster is ten million times stronger than the expansionary forces that may be causing the universe to expand, if indeed that is the case, a point we return to in later chapters.

If we write Hubble's law as

$$V/D = (\text{increment in D}/\text{increment in time})/D = H$$
$$= (\text{increment in D}/D)/\text{increment in time} = H$$

where H is Hubble's constant (it is not actually constant over time), D is the distance of an object away from us, and V is its recession velocity, then choosing H to be the reciprocal of thirteen billion years, we have for an increment in time of one year an expansion fraction of 7.7×10^{-11}.

Chapter 3

Then the Earth-Moon distance is about 3.83 x 10^8 metres, so this increases by the product of these last two quantities, or 3 cm per year. Whilst this may seem small, the same rate of expansion applied an object one light year away (9.46 trillion km) gives a movement of 72.8 km/year.

This might not seem much, but the constellation spot Hydra, which is some 2,000 M light years away from us, appears to be receding from us at 38,000 miles/second.

This assumes that the 'red shift' that Hubble's law is based upon is caused by expansion of the universe, and alternative explanations for redshift are given in Chapter 9.

Chapter 4

The Universe

> Now my own suspicion is that the Universe is not only queerer than we suppose, but queerer than we can suppose.
>
> J. B. S. Haldane, *Possible Worlds.*

The early universe

According to the Big Bang theory, during its first microsecond, the universe was gigantic homogeneous and isotropic quark-gluon plasma with extremely high energy density and very high temperature and pressure, which underwent a phase transition called a cosmic inflation to form protons and neutrons (three quarks each). During this transition, the universe expanded and cooled exponentially, and particle-antiparticle pairs of all kinds were rapidly created and destroyed in collisions at relativistic speeds and the excess of matter over antimatter of the universe began to develop.

The inhomogeneous nucleosynthesis model proposes that local pockets of mostly protons, and others of mostly neutrons, allowed more efficient nucleosynthesis reactions to form nuclei, resulting in more rapid increase in the density of the universe.

The recently detected Higgs boson, which imparts mass to fundamental particles such as quarks, probably played a major role in this early nucleosynthesis of the universe.

Chapter 4

With further cooling, the simplest atoms hydrogen and helium were able to form and, as cooling continued, hot clouds of these were able to cluster to form the first suns (Singh 2005).

Most theorists use the Friedmann-Lemaitre-Robertson-Walter (FLRW) model of the universe, date fitting to this leading to the conclusion that the universe is infinite and flat. Indeed, according to measurements by the Wilkinson Microwave Anisotropy Probe (WMAP), it is 99.6% certain that the universe is flat so that it can be modelled by three-dimensional flat Euclidean geometry.

Studying the stars

Stars are quasi-blackbodies so that they obey Wien's law for displacement of the wavelength of the light, they emit change in wavelength = $2.9 \times 10^{-6}/T$ nanometers where T is the absolute temperature of the object (°K). This law becomes less accurate as wavelength increases, and Max Planck's investigations of this inaccuracy contributed to his development of quantum theory.

The luminosity (L) or total energy emitted by a star can be calculated using the Stefan-Boltzmann law:

$$L = \text{Surface area} \times S \times T^4$$

where S is the Stefan-Boltzmann constant, which has the value 5.67×10^{-5} erg cm^{-2} sec^{-1} K^{-4}.

Distances to stars are determined by measuring their position from opposite sides of the Earth's orbit, that is, six months apart. The small artificial angular displacement observed relative to a background of very remote (essentially fixed) stars is called the parallactic angle.

Using the Earth's orbit as the baseline, the distance of the star is calculated from the parallactic angle, *p*:

$$\text{Sin}(p) = (\text{Distance earth to sun})/(\text{Distance Earth to star})$$

If *p* = 1" (one second of arc), the distance of the star is then 206,265 times the Earth's distance from the sun or 3.26 light years or one *parsec*.

The Universe

The nearest star, Proxima Centauri (a member of the triple system of Alpha Centauri), has a parallax of 0.76", meaning that its distance is 1/0.76 or 1.32 parsecs, or 4.3 light years. Stars at greater distances have even smaller parallaxes, and these are more difficult to measure directly. Hence, indirect methods mostly depending on the brightness of different types of star, are used to find the distances for the great majority of stars.

This is done by observing the apparent magnitude of a star's brightness, *m*, and its absolute magnitude M. Then the distance D of the star in parsecs is given by:

$$m - M = 5 \log(D) - 5$$

Types of stars

Plotting the nearest and brightest stars reveals four main types of star:

[1] *Main sequence.* Most stars fall into this category, one in which temperature and luminosity increase with mass over the fullest known range. Stars spend 90% of their life in this category. The sun is a relatively small star in this category.
[2] *Red giants.* These are very large and have low-range temperature and mid-range luminosity.
[3] *Supergiants.* These are relatively short-lived and extremely large stars with low-range temperature and high luminosity.
[4] *White dwarfs.* These are only 1% the size of the sun.

More than half the visible stars are *binary stars,* which revolve around their common centre of mass. There are also multiple star systems involving more than two stars.

Brown dwarfs are substellar objects too low in mass for hydrogen fusion, but some are thought to fuse deuterium or lithium. Some brown dwarfs have orbiting planets.

Chapter 4

Thermonuclear fusion in red giants

Like the sun, main-sequence stars produce energy from hydrogen fusion, as discussed in Chapter 3. In 2 or 3 solar mass stars this results in an excess of protons over electrons in their cores, eventually leading to helium fusion.

In normal cores, helium fusion increases quickly, increased gas and radiation pressure slowing fusion rates and contraction of the core. In smaller stars, the entire core must be compressed to degeneracy to reach the temperature required for helium fusion, and there is insufficient increase in radiation pressure to slow fusion rates and helium fusion occurs in a runaway fashion with explosions.

Eventually, however, all stars able to fuse helium become red giants by using the 'triple alpha process' in which each stage involves alpha particles or helium nuclei.

In the first stage, two alpha particles combine to form the very unstable isotope beryllium-8. This quickly fuses with another alpha particle to form a carbon-12 nucleus with the release of a gamma ray photon.

If the core is dense enough, a third step follows; in this, the carbon-12 nucleus combining with another alpha particle to form an oxygen-16 nucleus (Jones 2010).

Helium fusion gradually extends throughout the core, which is eventually converted to more compact inert carbon when shell hydrogen and helium fusion begin, increasing the star's size. In massive stars, this phase lasts only a few hundred thousand years, whereas for lower mass stars in the main group, it lasts for millions of years.

In the carbon fusion stage, carbon fuses with helium to produce oxygen, or with oxygen to form silicon. When carbon fusion extends to the outer core, the core collapses and fusion into heavier elements occurs in successively shorter phases. In each phase, the element being fused is exhausted and fusion ceases and the core collapses and heats up to begin the next fusion reaction.

These successive phases continue until the core becomes iron ash. As both fusion and fission of iron require energy input, with no energy source remaining to maintain its equilibrium, the outer layers of the star collapse into the iron core. If this is relatively small, the result is a neutron star maintained by the force of repulsion between neutrons (neutron degeneracy). If the core is more massive, the result is a black hole, which emits no light but has a strong gravitational field which may form a hot and radiant accretion disk, which helps identify the black hole.

Neutron stars have a solid surface and very strong magnetic fields. They are typically only some 20 km in diameter but have mass about twice that of the sun, so that their density is circa 100 trillion times that of water. Thus their internal pressure is too high for atoms to exist and protons and electrons are compacted together into neutrons. Pulsars are rapidly spinning neutron stars, which emit powerful, pulsating radio waves and more than 500 pulsars have been identified in the Milky Way.

Stars too small to become neutron stars or black holes become white dwarfs. Having exhausted all their nuclear fuel, they collapse and become extremely dense, typically containing the mass of the sun in the volume of the Earth. Over billions of years, they cool to become inert remnants called black dwarfs.

Variable stars

All stars vary in brightness slightly, but some stars vary greatly, some with great regularity, some seemingly randomly. Cepheid variables are bright yellow supergiant stars that pulsate regularly, their periods varying from 1.5 days to more than 50 days, their brightness being hundreds of times more than that of the sun and varying in proportion to their periods, this allowing estimation of their distance.

The most spectacular variable star is the so-called temporary star, or nova. Novas may be up to 200,000 times as bright as the sun, achieving this by rapid fusion reactions, which blow of debris at speeds up to 960 km/sec. Some novas repeat this process periodically until they lose too much mass to continue.

Supernovas are caused by fusion in a degenerate star or collapse of the core of a massive star when fusion becomes unable to sustain the core against its own gravity. A degenerate white dwarf may accumulate sufficient material from another white dwarf by accretion or by merger, raising its core temperature sufficiently to initiate runaway carbon fusion.

Chapter 4

When supernovas explode, they sometimes brighten for a few days to ten billion times the sun's brightness before fading away permanently. They leave behind expanding wreckage seen as bright gaseous clouds, or nebulas.

An example is the Crab nebula, first observed from earth as a supernova in 1054. Sometimes, a pulsar is also left as a remnant in the centre of the wreckage.

Supernovas play a significant part in enriching the interstellar medium with higher mass elements and their expanding shock waves can trigger the formation of new stars. Studies of supernova remnants indicate that they occur about three times per century in the Milky Way.

Rogue planets

It is estimated that there might be 100,000 times more rogue planets than stars in our Milky Way galaxy. These may have formed alone early on in the universe or have been ejected from a solar system.

The latest rogue planet found, CFBDSIR2149, is about 100 light years away and, at present at least, it seems loosely associated with a moving group of stars called the AB Doradus moving group, the closest such group to our solar system.

Other galaxies

In a solid rotating object, the mass distribution is uniform. A flat disc, for example, has the same mass/unit area throughout. Because the sun contains nearly all the mass of the solar system, however, both the mass and emission distribution curves for our solar system take the exponential form shown in Figure 4.1.

Galaxies are more like the solar system than a solid body because they consist of relatively small objects moving around a gravitational potential well (the galactic centre). The mass distribution of a galaxy can be inferred by mapping the light intensity of its emissions.

To help explain the mass distributions observed in some galaxies, 'dark matter' was postulated, and this is believed to make up about most of the mass of the universe. Dark matter is described as 'cold' because it has low energy, cannot radiate photons, and only interacts with other particles through gravity and perhaps a weak force similar to that between neutrinos and antineutrinos.

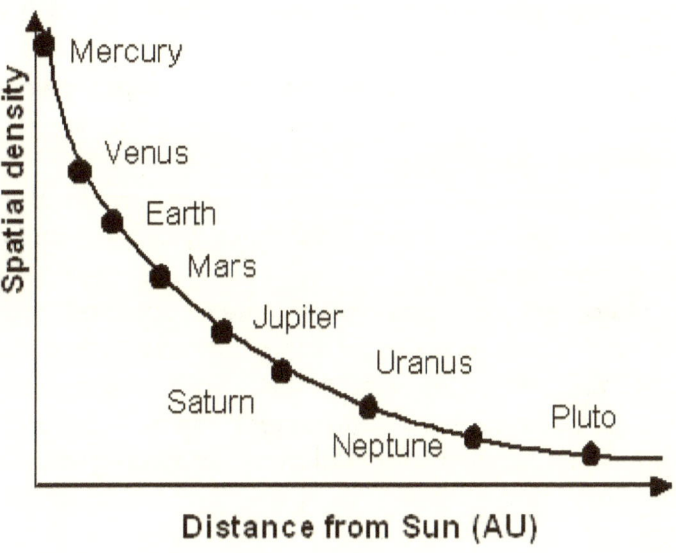

Figure 4.1. Mass distribution in the solar system.

NASA's Galaxy Evolution Explorer found what appeared to be three dozen massive galaxies in Earth's corner of the universe that may be as little as 0.1 to 1 billion years old, our Milky Way being 10 billion years old.

The new galaxies emit about ten times more UV than the Milky Way, indicative of their recency.

New stars and planets

Stars eventually burn out as supernovas, releasing masses of material into space. Sometimes this matter is able to accumulate sufficiently and condense to form a new star.

Several new stars are thought to have formed in part of the Eagle Nebula, 7000 light years away.

Chapter 4

In 1995, new stars were reported to be emerging from dense structures of gas in the Orion and Eagle Nebulae, thousands of light years away.

The star 51 Pegasi is forty-five light years away and its light signal suggests a planet 190 times larger than the earth and twenty-two times closer to its star than is Earth, but it is doubtful that a planet could form so close to a star.

A planet 2500 times heavier than Earth (eight times heavier than Jupiter) is thought to be twice as close to the star 70 Virginia as Earth is to the sun.

A wobble in the star 47 Ursae Majoris in the Big Dipper constellation suggests a planet 1000 times heavier than Earth and twice as far from its sun.

Two streamers of gas have been detected connecting the inner and outer disks of cosmic material surrounding the star HD142527, which is about 450 light years from earth. Astronomers believe these streamers were caused by the gravitational pull of two giant planets, providing a mechanism by which stars can continue to grow by accumulating cosmic material.

Quasars

Quasars or 'quasi-stellar radio sources' were first detected by radio astronomers in the 1950s, and their spectra were first seen by Maarten Schmidt of the Hale Observatories in Pasadena, California, in 1963 (Hoyle 1983). Schmidt discovered that the emission lines in the spectrum of quasar 3C 273 were known lines that exhibited a far stronger redshift than in any other known object.

Redshift can be caused by the Doppler effect, which shifts the wavelength of light from celestial objects toward the longer red wavelengths when the objects are moving away from the observer. Such redshift is called cosmological, and the recession velocity can be calculated from its magnitude from Hubble's law according to which recession velocity is proportional to the distance of the object. This calculation indicates that quasar 3C 273 is 1.5 billion light years from the earth.

Several thousand quasars had been identified by the end of the 1980s and the redshifts of many of them determined, in a few cases the shift factor being greater than 4, suggesting a velocity greater than 93% of that of light at a distance greater than 10 billion light years.

In the 1990s astronomers discovered quasars that are surrounded by a haze of visible light from the quasar's host galaxy. Judging from the energy received on earth from such distant objects, some quasars produce more energy than 2000 ordinary galaxies combined (*Encarta* 1999).

The emissions of many quasars are variable, some of these being estimated to be only one light year wide.

Astronomers believe that quasars are supermassive black holes at the centre of an elliptical galaxy and surrounded by swirling matter. The matter emits the quasar's radiation as it falls into the black hole. Many quasars are millions of times more massive than the sun but occupy about the same space as our solar system.

Black holes

Black holes are believed to form when giant stars reach the end of their life with a supernova explosion, which collapses their core, leaving a dense stellar remnant. Normally, the result is a neutron star, but if the star's core weighs more than a few solar masses, its collapse is so powerful that the neutrons are broken into their constituent quarks. The core is then believed to become an infinitely dense singularity sealed off from the universe by the so-called 'event horizon', which prevents radiation from escaping and traps objects that enter it.

Black holes are detected by observing the gravitational effects they have on nearby stars and gigantic "supermassive" black holes with the mass of millions of suns have been found at the heart of many galaxies, including the Milky Way (Sparrow 2007).

In 1994 astronomers used the Hubble Space Telescope to find the first convincing evidence of a black hole. They measured the acceleration of gases around the centre of the galaxy M87 and concluded that an object of about three billion solar masses must be present. Several similar discoveries have been made since (*Encarta* 1999).

Chapter 4

Dark matter

The amount of matter in the universe dictates its gravity, which in turn could decide its fate. Visible light reveals hot objects such as stars, and infrared radiation reveals dark clouds of gas and dust. Dark matter, however, can only be detected by its gravity, but is believed to be four times the density of normal matter and comprise up to 90% of the mass of the universe (Sparrow 2006).

There two types of dark matter, MACHOs (massive compact halo objects), which probably comprise less than 20% of all dark matter, and WIMPS (weakly interacting massive particles) and the 'grand unified theories' predict such particles, axions, photinos, and zinos (Gribbin 1998).

MACHOs include stray planets, black dwarfs and black holes, and these may be found in the apparently empty 'halos' around spiral galaxies such as our own. Several black holes have been detected in our galaxy's halo by their distortion of light from distant galaxies.

A sequence of star images from the Magellanic Cloud, a galaxy near ours, showed gravitational lensing that temporarily bent and greatly brightened the light of a minor star, and this was compelling evidence of the temporary presence of dark matter between the star and Earth (Sparrow 2006).

WIMPs are particles that do not interact with normal matter and include neutrinos, particles with only tiny mass which are radiated by stars throughout the universe as a result of nuclear reaction in which protons decay into a neutron, a positron (anti-electron), and an unstable anti-electron neutrino that quickly becomes an electron-neutrino.

Fewer solar neutrinos have been detected than were expected; one tentative explanation being that they interact with matter and become heavier tau or muon neutrinos.

Dark matter is described as 'cold' because it has low energy, cannot radiate photons, and only interacts with other particles through gravity and perhaps a weak force similar to that between neutrinos and antineutrinos.

Conclusion

The universe is massive, indeed seemingly infinite, and is filled with a great variety of objects, including the following:

[1] Galaxies. Many 'new' galaxies have been discovered. In 2003 the Hubble Ultra-Deep Field (HUDF) used an upgraded camera to obtain pictures indicating some 10,000 galaxies up to thirteen billion light years away in the constellation Fornax.

In 2009 the Great Observatories Origins Deep Survey (GOODS) found evidence in small areas of the constellation Fornax of 7,500 galaxies in an area of our skies only about the size of the full moon.

[2] Various types of stars, including main sequence (the most common and including our sun), red giants, supergiants, binary stars, neutron stars, and white dwarfs.
[3] Variable stars including cepheid variable, novae, and supernovas.
[4] Many 'new' stars and planets, including rogue planets, have been discovered, our solar system now being deemed to have only eight planets, but also five dwarf planets.
[5] Quasars and black holes.
[6] Dark matter and dark energy.

The latter was postulated to account for the discovery that, according to Hubble's law, the rate of expansion of the universe was increasing, suggesting that dark energy had grown stronger than gravity. The issue of whether the universe is really expanding will be discussed in Chapter 9.

Chapter 5

The Interstellar Medium

> *The universe is wider than our views of it.*
> Henry David Thoreau, *Walden*.

Interstellar medium feedback processes

The interstellar medium (ISM) contains molecular and atomic gas (99%) as well as dust (1%). Most of the ISM is atomic hydrogen, the next most plentiful ingredients being molecular hydrogen gas then other gases and then dust. Most of the dust is found in molecular clouds and is carbon based, but there is some evidence of silicon-based dust.

The universe recycles much of the material in supergiant stars when supernova explosions return the material the stars used during their existence to the interstellar medium. The expelled planetary nebulae and supernova remnants then eventually contribute to the formation of new stars.

Most stars also have small but regular eruptions and thus lose mass to the ISM throughout their lifetimes. The most massive stars are the best recyclers, and these return almost 90% of the materials they use during their lifetimes to the ISM as supernova remnants (Jones 2010).

The ISM is about 5% of the mass of the universe, and about 10% of the mass of the galaxy (*Oxford Interactive Encyclopedia* 1977).

Regions of the interstellar medium

The interstellar medium can be divided into three types of region based upon their temperature (called the phase of the ISM):

(a) Cold (10s °K), mostly molecular gases.
(b) Warm (100-1000s °K), mostly neutral hydrogen atoms.
(c) Hot (millions °K), mostly ionized (HII regions).

Electromagnetic emissions from colder regions are mainly in the longer radio and microwave wavelengths.

The coldest regions are dark nebula, some of which are found from star counts, for example the Coalsack Nebula. Others, such as the Horsehead Nebula, are revealed by emissions from behind them which heat their gases. The Horsehead Nebula is several tens of parsecs across, large enough to contain our solar system and the stars surrounding it.

Often, dense concentrations of gas and dust called molecular clouds are at the centres of dark nebula. Their density and turbulence sometimes make large areas of cloud collapse to form protostars. As the new stars evolve, they heat leftover gas, and the molecular cloud transforms into a HII region such as the Orion Nebula.

The interplanetary medium

The interplanetary medium (IPM) contains interplanetary dust, cosmic rays (mainly protons and alpha particles), and hot plasma from the solar wind which is mostly made up of hydrogen ions and electrons. The IPM fills the *heliosphere*, the teardrop-shaped region of space containing the solar system, which moves through space at a speed of about 25 km/sec.

Temperatures of dust particles in the asteroid belt range from 200 °K down to 165 °K. Near the sun, the density of the IPM is 100 particles/cc, this reducing by inverse square law with distance from the sun to about 5 particles/cc near the Earth.

Scattering of light from dust in the IPM causes a number of visible optical effects, particularly at sunrise and sunset.

Chapter 5

Being a plasma, the IPM carries the sun's magnetic field and is highly electrically conductive. As a result, the sun's magnetic field near Earth is 100 times what it would be if space were a vacuum.

The solar wind leaves the sun at speeds of 250-1000 km/sec and impacts directly on bodies with no magnetic field such as the moon, resulting in emission of faint X-ray radiation. The magnetospheres surrounding planets with their own magnetic fields such as the Earth and Jupiter largely protect them from solar wind, but material from solar wind can sometimes penetrate the magnetosphere, causing aurorae and populating the Van Allen belts with ionized material.

The boundary of the IPM and the solar wind is the outer edge of the solar system, and this is at 110-160 AU from the sun and is called the heliopause.

The magnetic field of the heliosphere prevents most particles outside it from entering, exceptions being cosmic dust grains, cosmic rays, and 'interstellar neutrals', neutral atoms from the interstellar medium some of which eventually become cosmic rays.

Composition of the interstellar medium

The interstellar medium has multiple phases: atomic, ionic, or molecular. The thermal pressures of these phases are in approximate equilibrium. Magnetic fields and turbulent motions also produce pressure in the ISM, and these have more effect than the thermal pressure.

Cool, dense regions are mostly molecular with densities of up to a million molecules/cc (water has a density of 10^{22} molecules/cc). Hot, diffuse regions are mostly ionized and have densities as low as 10^{-4} ions/cc.

As with the IPM, 99% of the ISM is gas in any form, and 1% dust. The gas in the ISM is 89% hydrogen and 9% helium from primordial nucleosynthesis, and 2% heavier elements referred to as 'metals', which were provided by stellar evolutionary processes such as supernovas.

Table 5.1 shows the properties of the regions of the Milky Way, the hot IM region being coronal gas, which also includes highly ionized 'metals' (*Wikipedia* 2014).

The Interstellar Medium

Table 5.1. Properties of regions of the Milky Way.

Region	Volume (%)	Scale height (parsec)	Temp. (°K)	Density (atoms/cc)	State of Hydrogen
Molecular clouds	<1	80	10-20	100-10^6	molecular
Cold NM	1-5	100-300	50-100	20-50	neutral atomic
Warm NM	10-20	300-400	6,000-10,000	0.2-0.5	neutral atomic
Warm IM	20-50	1000	8000	0.2-0.5	ionized
Hydrogen stage II	<1	70	8000	10-100	ionized
Hot IM	30-70	1000-3000	10^5-10^7	.01-10^{-4}	ionized
NM = neutral medium, IM = ionized medium					

A 'three-phase equilibrium' model of the ISM has a cold condensed phase (T < 300 °K) of clouds of neutral and molecular hydrogen, warm intercloud phase (T circa 10,000 °K) of rarefied neutral and ionized gas, and a very hot phase (T circa a million °K) shock heated by supernovas and making up most of the volume of the ISM.

Stars form in the densest regions of the ISM, molecular clouds, and replenish the ISM through planetary nebulae, stellar winds, and supernovas. Matter exchange between stars and the ISM plays a part in determining the rate at which a galaxy depletes its gaseous content, and therefore how long it continues to form stars (*Wikipedia* 2014).

Neutral atomic interstellar gas

Neutral atomic interstellar gas is mainly hydrogen, each hydrogen atom having one proton and one electron. In the Bohr model of atomic structure, electrons orbit in various discrete circular orbits around the nucleus. In reality, the electrons have quasi-chaotic orbits, nevertheless, they can only have discrete energy states.

Chapter 5

In neutral hydrogen, the electron is at its lowest energy state, or *ground state*. Neutral atomic hydrogen has two ground states, the lower when the proton and electron have opposite spin, the higher when they have the same spin. Transition from the higher to the lower state is called *spin-flip transition*, which involves a change in energy equivalent to a photon with wavelength 21 cm and thus neutral atomic hydrogen emits electromagnetic radiation at this wavelength when excited. Observation of this radiation from other galaxies is used to determine their mass distributions (Jones 2010).

Ionized atomic interstellar gas

Ionized atomic interstellar gas is mainly hydrogen, but also includes significant amounts of helium, neon, nitrogen, oxygen, carbon, sulphur, and iron.

An atom is ionized when one or more electrons are removed from the neutral atom. In space, ionization is usually caused by electromagnetic radiation, which is highly energetic near the hottest stars.

When gases are moving very fast, mechanical processes can cause ionization; for example, a supernova event can produce an expanding shell of ionized gas. Near a black hole, gas may be ionized by high-speed collisions between gas particles and shear between clouds of gas in an accretion disk being consumed by the black hole.

Study of UV, IR, and MW radiation enables detailed study of ionized interstellar gas (Jones 2010).

Molecular interstellar gas

Interstellar gas in molecular form is mostly molecular hydrogen, H_2. The next most plentiful gas is carbon monoxide, CO.

The H_2 molecule has no dipole moment and thus no preferred orientation in the presence of electric, magnetic, or gravitational fields. Thus these fields cannot cause electrons to change energy state, and no electromagnetic radiation will be emitted. Thus emission from hydrogen molecules only occurs in extreme situations such as in black hole accretion disks or supernova events.

Since CO is always associated with molecular hydrogen, astronomers use low energy, long wavelength CO radio emissions to study H_2 in space (Jones 2010).

Interstellar dust

The location and nature of dust can be determined from light reflected and scattered by it while its blackbody spectrum determines its temperature.

Interstellar dust ranges from large molecules to small carbon or silicate dust particles up to about half a millimetre in size. Long molecular chains of polycyclic aromatic hydrocarbons (PAHs) are the smallest particles of dust, and large dust grains may be covered with water ice.

On Earth, PAHs are considered air pollution, and the Centers for Disease Control describe PAHs as "chemicals that are formed during the incomplete burning of coal, oil, gas, garbage, or other organic substances."

PAHs are most prominent in 'star-forming regions' of the interstellar medium where it might be expected that electromagnetic radiation levels would be high enough to ionize and dissociate PAH molecules.

Mid-infrared radiation (5-50 micrometres) is dominated by PAH emissions, whilst near-infrared (1-5 micrometres) is primarily from atomic gas ionized by radiation and collision, and far-infrared (50-200 micrometres) is mainly blackbody radiation from rare, highly ionized atomic and molecular gases (Jones 2010).

According to one estimate, up to 40,000 tons of cosmic dust reaches the Earth's surface every year.

Cosmic dust plays a key role in star formation and mass loss when a star is ending its life. In our solar system, dust plays a major role in comets, zodiacal light, Saturn's B ring spokes, and the diffuse outer planetary rings of Jupiter, Saturn, Uranus, and Neptune.

Most of the interplanetary dust that falls on Earth comes from meteoroids and is of diameter 0.05 to 0.5 mm and average density 2 g/cc with about 40% porosity.

Circumstellar dust contains CO, silicon carbide, amorphous silicate, PAHs, polyformaldehyde, and water ice.

Chapter 5

Asteroidal dust is carbonaceous and resembles that from meteorites, and cometary dust resembles interstellar grains and contains silicates, PAHs, and water ice.

The interstellar medium also contains radioactive aluminium-26 thought to have originated from supernova explosions and nucleosynthesis in aging giant stars. Supernova condensates are also abundant in calcium-44, suggesting that they condensed containing radioactive titanium-44, which has only a sixty-five-year half-life.

Stardust

Stardust is contained in meteorites and originated from the refractory dust that condensed from clouds of cooling gases from which the solar system formed. Its main constituents are graphite, silicon carbide, aluminium oxide, and magnesium oxide, the latter two compounds being combined as spinel, a glassy substance often used as gemstones. Supernova condensates are also rich in calcium.

Stardust is only 0.1% of the total mass of interstellar solids, but its presence verifies that the solar system was condensed from hot gas, so hot as to vaporize any solids.

Interstellar extinction and reddening

Detailed studies of the clusters at various distances from us showed that the light from remote stars is reduced in intensity (interstellar extinction) and reddened (interstellar reddening) as it passes through the sparse material of the interstellar medium. Reddening occurs because the shorter blue wavelength photons are more easily absorbed or scattered than the red ones, this being the reason why the sky is blue during the day and reddens when close to the horizon at sunset.

The interstellar extinction curve was obtained by observing the spectra of hundreds of thousands of stars and shows how different wavelengths of light are extinguished (absorbed, scattered, and reflected) as they pass through the ISM. This curve shows a gradual increase in extinction with decreasing wavelength, and also a 'graphite bump' at 217.5 nanometers caused by graphite and silicate grains.

The extinction curves for other galaxies differ; two of the galaxies closest to ours not having the graphite bump, presumably because they do not contain as much carbon based dust. The curve for our galaxy is, however, very similar to those of the most actively 'starbursting galaxies' with a high rate of star formation.

Interstellar heating and cooling

Interstellar gases are heated by the following:

(a) Low-energy cosmic rays are far more prevalent than high-energy cosmic rays and cause excitation and ionization of gases.
(b) Photoelectric heating. The UV radiation from hot stars can energize electrons in the smallest dust grains so that they are ejected. This heating mechanism dominates in HII regions, but is negligible in diffuse areas of the ISM.
(c) X-rays can ionize atoms and cause secondary ionization in ions. As the X-ray intensity is usually low, X-ray heating is only efficient in warm, less dense media.
(d) Chemical heating when a pair of hydrogen atoms in a grain combine to form molecular hydrogen.
(e) Grain-gas heating is caused by collisions between gas atoms and molecules at high density with dust grains. Grain heating is significant in supernova remnants in which densities and temperatures are high. Gas heating is dominant in high density molecular clouds.

In most regions of the ISM, except hot regions and molecular clouds, gases are cooled by photon emission after inter-particle collisions excite electrons to higher energy levels.

Bacteria in the ISM?

Fred Hoyle and Chandra Wickramasinghe began studying the composition of interstellar dust in the 1960s.

In 1968, polycyclic aromatic molecules were detected in interstellar dust, and in 1972, convincing evidence that the dust contained porphyrins was obtained.

Chapter 5

In 1974, Wickramasinghe proved that molecules of poly-formaldehyde, which is closely related to cellulose, were in the ISM.

In 1979, Hoyle and Wickramasinghe compared the interstellar extinction curve with that for dried bacteria, obtaining a very close match (Hoyle and Wickramasinghe 1985). This curve has a broad 'knee' centred at about 2.3 wavelengths per micrometre, a feature that could be explained if the ISM dust was a certain size and translucent. Dried bacteria have the appropriate size range and refract light as irregular hollow spheres, so Hoyle and Wickramasinghe concluded that the extinction curve was probably caused by dried, frozen bacteria.

This result has not been widely accepted, but Hoyle (1983) also pointed out that such bacteria could be responsible for the cosmic microwave background (CMB) radiation usually attributed now to the Big Bang.

Indeed, some bacterial species exist inside small needle-like sheaths, which conserve water in the harsh medium of space, the needle shape being ideal for transforming starlight into radio waves.

That bacteria can survive in space was spectacularly demonstrated when in 1969 the *Apollo 12* astronauts recovered a TV camera that had been left on the moon by an unmanned lunar lander two and a half years earlier. When it was examined back on Earth, *Streptococcus mitis* were found still alive on it.

Survival in such harsh conditions with massive temperature swings and no water is possible because bacteria can become 'micro-cannibals', most bacteria dying to supply the rest with protective proteins, gums, sugars and cryoprotective substances. For example, bacteria seem to have survived 4,800 years in the brickwork of Peruvian pyramids and as much as 300 million years in coal.

The ultimate survival mechanism for bacteria, however, is formation of spores with protective coats that become completely dormant.

Some bacteria form endospores, the original cell replicating its chromosome and surrounding one copy with a durable coating which can survive in boiling water for long periods of time.

The Interstellar Medium

Hoyle (1983) also notes that, when carbon vapour is cooled gradually, not rapidly as with vapour from a fire, long slender needles of carbon are formed; and these are also the right shape to convert starlight into radio waves, providing another alternative explanation for the CMB.

Conclusions

The ISM has many different types of regions and plays an important role in the universe, particularly in formation of new stars from aggregations of cosmic dust and supernova remnants.

Of particular note, however, is the possible presence of bacteria in the ISM and that carbon whiskers formed by slow cooling of supernova remnants may be responsible for the cosmic microwave background radiation that is normally attributed to expansion of the universe since the Big Bang, a matter which will be discussed further in Chapter 11.

Chapter 6
The Theory of Relativity

> *In order to give physical significance to the concept of time, processes of some kind are required which enable relations to be established between different places. It is immaterial what kind of processes one chooses for such a definition of time. It is advantageous, however, for the theory, to choose only those processes concerning which we know something certain. This holds for the propagation of light 'in vacuo' in a higher degree than for any other process which could be considered, thanks to the investigations of Maxwell and H.A. Lorentz.*
>
> Albert Einstein, *The Meaning of Relativity.*

The Michelson-Morley experiment

In 1887, Albert Michelson and Edward Morley attempted to measure the speed of the light with respect to the aether. They used an interferometer, which split light in two directions, one in line with the motion of the earth, the other perpendicular to it. As no difference in the speeds of the two beams was found, the aether was proven to not exist.

The experiment also proved that the speed of light is the same for all observers, regardless of their relative motion.

The Theory of Relativity

This finding motivated Einstein's development of the special theory of relativity eighteen years later, choosing the speed of light in a vacuum as his fundamental constant, as noted in the quotation above.

The Lorentz transformation

It was Irish physicist George Francis Fitzgerald who first proposed in 1889 that an object would shorten in its direction of motion relative to an observer. Hendrik Lorentz subsequently developed the Lorentz transformation relating two coordinate systems in relative motion (Prokhovnik 1967):

$$x = (x' + ut')/F, \ y = y', \ z = z', \ t = (t' + ux'/c^2)/F$$

where $F = \sqrt{1 - u^2/c^2}$, u is the velocity of the object, c is the velocity of light, and F is the Lorentz factor. Experiments failed to find the Fitzgerald contraction, however, but the Lorentz factor was the basis for Einstein's special theory of relativity.

The special theory of relativity

The Michelson-Morley experiment had failed to measure the velocity of light relative to the aether; the two fundamental postulates of Einstein's special theory of relativity (STR) were the following (Prokhovnik 1967):

1. The laws of nature are equally valid for all inertial frames of reference.
2. The velocity of light is invariant for all inertial systems, being independent of the velocity of its source and being constant for all observers.

Chapter 6

The basic equations of the special theory of relativity are proved as follows (Eddington 1924):

In 4-D space we allow a point (x,y,z,T) to move to (x+dx,y+dy,z+dz,T+dT) and write the square of the distance moved in the quadratic form:

$$ds^2 = g_x dx^2 + g_y dy^2 + g_z dz^2 + g_T dT^2 + 2g_{xy} dxdy \; \text{-} \; \text{-}$$

where T is a fourth dimension with the same spatial units as x,y,z, and the coefficients g_x etc. are functions of x,y,z,T and there are six 'cross' or g_{xy} type terms.

Using rectangular coordinates the 'cross' terms in this equation vanish (equivalent to taking a vector dot product) and the coefficients g_x etc. reduce to unity so that we have

$$ds^2 = dx^2 + dy^2 + dz^2 + dT^2$$

Using complex numbers for time t we here let dT = icdt where c is the velocity of light and i = sqrt(-1), giving

$$ds^2 = dx^2 + dy^2 + dz^2 - c^2 dt^2$$

Here, 'i' yields the 'imaginary' part of a complex number. Complex numbers are often used to represent electromagnetic wave propagation, the 'real' part of the complex number representing the mean intensity, and the imaginary part representing the amplitude of the usually sinusoidal variation of electromagnetic wave intensity with time (Kreyszig 1979).

Whether this artifice is justified as a means of obtaining the time dilation formula is very doubtful, and in fact, the fourth time dimension should be brought into play as a result of the coordinates (x,y,z) varying according to the components of velocity in these directions.

Now consider a clock moving from (0,0,0) to (0,y+dy,0). The reading of the clock will be proportional to the distance travelled in space-time, which is calculated by integrating

$$-ds^2 = c^2 dt^2 - dy^2 = c^2 dt^2 [1 - (dy/dt)^2/c^2]$$

so that denoting the speed dy/dt = v, the difference in the clock readings in the moving (S*) and stationary system (S), as observed by an electromagnetic signal between them, will be proportional to ds/c, or

$$[t_2' - t_1'] = \text{Integral } [dt(1 - v^2/c^2)]^{1/2} = [t_2 - t_1] F$$

where $F = (1 - v^2/c^2)^{1/2}$ is the Lorentz factor.

Hence the moving clock appears slower compared to the clock in the stationary frame S, a phenomenon called *time dilation*, but only because of the time taken by an electromagnetic signal to relate the two times.

Recriprocally, the co-ordinates of the clock in the stationary system S are (0,0,0,t) and relative to the observer in the moving system S* they are (-vt',0,0,t'). According to the observer in S*, therefore, a clock stationary in S also appears to run slowly according to the reciprocal relation (Prokhovnik 1967):

$$[t_2 - t_1] = [t_2' - t_1'] F$$

According to the special theory of relativity length and mass also alter in this moving (relative to us) system

$$L' = L(1 - v^2/c^2)^{1/2}, \quad m' = m/(1 - v^2/c^2)^{1/2}$$

and the contraction in length observed is called the Lorentz-Fitzgerald contraction, whilst m = the rest mass and m' = the relative mass. The very doubtful proof of the latter result involves arbitrarily replacing ds^2 with $-ds^2$ in the foregoing equations (Eddington 1924). Indeed, Einstein said, *"It is better to introduce no other mass concept than the 'rest mass' m. Instead of introducing M [here m'] it is better to mention the expression for the momentum and energy of a body in motion."*

Einstein's general theory of relativity (GTR) included the effect of gravity on the foregoing equations, particularly the bending of 'space-time' by massive objects (Einstein 1922). Like the special theory, it does not have great practical applications and its main relevance is to astronomy and astrophysics.

Chapter 6

E = mc²

This is the most famous equation in history, but it is only an approximation, and a crude one at that. When the units are such that c =1, STR gives

$$m' = m(1 - v^2)^{-1/2}$$

$= m + \frac{1}{2} mv^2$ approximately if the speed is small compared with that of light. The second term is the kinetic energy, and the first the 'internal energy' of the body.

If we denote the total energy as E and remove the restriction c =1 we obtain (Eddington 1924):

$$m' = E/c^2$$

The mathematics of this proof, as with most derivations in special relativity, is poor, but the result is easily envisaged if we imagine photons of light being instantly propagated from some excited source with the speed of light. Though generally deemed massless, they have been found to have a tiny mass under certain conditions and, of course, they travel at the speed of light, their kinetic energy not having the usual ½ term because they were not accelerated from rest but instantaneously propagated from their source. As often happens, therefore, nuclear reactions, here those producing light, change the rules somewhat.

The result of this famous equation, of course, is that the principles of conservation of mass and of energy are fused, often termed an 'equivalence of mass and energy'.

Thus when an electron and positron collide and annihilate one another, when the particle masses are lost, the total energy is conserved (perhaps only approximately) in the resulting electromagnetic waves that are produced, the energy of a photon being calculated as:

$$E = hf$$

The Theory of Relativity

where h is Planck's constant and f is the frequency of light (vibrations/sec). Under certain conditions, such as those in nuclear reactors, photons have a small mass.

Deflection of light by the sun

The general theory of relativity takes into account gravitational effects; it's most famous prediction being deflection of light around the sun. On May 29, 1919, a rare solar eclipse occurred and two groups of scientists went to Sebral in northern Brazil, and to the Isle of Principe in the Gulf of Guinea, to measure the displacement of the sun's light and compare it to Einstein's prediction of 1.74 seconds.

The Principe measurement was 1.61 seconds of arc, with a margin of error of 30 seconds, whilst the Sobral measurement was 1.98 seconds of arc with a margin of error of 18 seconds. These results were greeted with enthusiasm by supporters of the theory of relativity as proof of it, and Einstein became a celebrity overnight.

Subsequent eclipse measurements gave the following results (margin of error shown in brackets):

Greenwich, Australia, Sept. 21, 1922	1.77 (0.40)
Potsdam, Sumatra, Sept. 21, 1929	1.82 (0.20)
Sternberg, USSR, June 19, 1936	2.73 (0.31)
Sendai, Japan, June 19, 1947	2.13 (1.15)
Yerkes, Brazil, May 20, 1952	2.01 (0.27)
Yerkes, Sudan, Feb. 25, 1952	1.70 (0.10)

There is considerable variation in the results, however, and these results alone are insufficient to be convincing proof of the theory of relativity (Bernstein 1973).

Proof of time dilation?

In 1971, Hafele and Keating placed caesium clocks on two aeroplanes, flying in opposite directions around the equator back to their starting points. They calculated that because of the time dilation of STR, and the gravitational effects of GTR, the eastward clock should loose 40 nanoseconds, and the westward clock should gain 275 nanoseconds. The measured results were respectively 59 and 273 nanoseconds.

This seems a good result, but it transpires that Hafele and Keating made a number of "corrections" to their data to average out the errors between the clocks used, and that variations between clocks moving in similar directions were large enough to invalidate the results (Burchell 2013).

As noted in Chapter 15, however, the Hafele and Keating clock experiment was repeated in 1971, and the theory of relativity predictions verified to within 4% accuracy.

Global positioning system (GPS) satellites transmit a time signal to a GPS receiver, which calculates the distance from the satellite by measuring the time the signal takes to reach the receiver.

GPS satellites carry accurate atomic clocks which run at a different speed to that on Earth because of time dilation (STR), which should slow the clocks by 7,200 nanoseconds/day, and reduced gravity (GTR), which should speed up the clocks by 45,900 nanoseconds/day, giving a net gain of 38,700 ns/day.

Thus GPS engineers adjust the clock rates by about 38,500 ns/day. According to Burchell (2013), this adjustment makes no difference to GPS accuracy. Indeed, according to Stormfront.org (2014) Ronald R. Hatch, director of Navigation Systems at NavCom Technology wrote:

In principle, the critics of GPS in the relativity debate have not been completely wrong. The neglected 7 factor could hurt us. The OCS software should be reformulated. Nevertheless, in practice, neglect of relativity does not contribute measurably to the GPS error budget, as the OCS software is currently configured.

Relativity and particle accelerators

According to JK Cannizzo, an astrophysicist at NASA:

The strongest direct evidence [of STR] comes probably from particle accelerators, in which subatomic particles such as electrons and positrons are accelerated to within a few inches per second of the speed of light. We can observe very clearly and

The Theory of Relativity

accurately the changes in, for instance, the apparent masses of the particles. They are observed to increase dramatically, and in fact new and much heavier particles can be created by making counter-rotating beams of, say, electrons and positrons, collide head-on with each other. Special relativity has played a key role in the design and operation of particle accelerators for many years (Ask an Astrophysicist website, 2014).

Such observations are based on the mass-energy equivalence implied by $E = mc^2$, but arguably, reactions between subatomic particles cannot be expected to obey the simple laws of Newtonian mechanics.

Thus the behaviour of colliding heavy ions can only be explained if increases in density as a result of Lorentz contraction are considered. Such contraction also increases the intensity of the Coulomb field perpendicular to the direction of motion, and this effect has been observed.

CPT symmetry (of charge conjugation, parity transformation, and time reversal) is not preserved in some physical phenomena and the standard model extension (SME) is a field theory that combines the standard model of the atom discussed in Chapter 1 with general relativity, and operators that break and preserve CPT symmetry.

In the presence of fixed background fields, for example, observer and particle transformations are not equivalent but inverse, and symmetry is lost.

Thus, amongst the many complex interactions that occur between subatomic particles, photons, and gravitons, fundamental laws of macroscopic physics, including those of STR and GTR, are often broken.

Travelling faster than light

Recently, scientists working on the Oscillation Project with Emulsion-tracking Apparatus (OPERA) at CERN sent neutrinos through the Earth to the Gran Sasso laboratory, 730 km away. They completed the journey 60 nanoseconds quicker than if they had been travelling at the speed of light, for which the travel time would have been 2.4 milliseconds, so that the neutrinos went 6 km/sec faster than the speed of light (299,798 km/sec).

Chapter 6

Despite this experiment having been repeated many times, however, some doubt remains about this finding.

Radio waves from a pulsar, however, have been found to be travelling faster than the speed of light.

Photons with different speeds

It has recently been reported that astronomers observing distant objects in the universe have detected photons with speeds that differ by about a second/billion years, showing that the speed of light is not constant but, to some extent at least, depends upon the source.

Recently, a telescope viewing a supernova from over sixteen billion light years away clocked a low energy photon arriving five to seven seconds later than its high energy equivalent. This, may have been because of different ejection speeds or interference at the source, or more likely because of differing levels of interaction with the interstellar medium.

Not for the first time, however, the fundamental premise of the theory of relativity, namely that the speed of light is constant and cannot be exceeded, was found to be false.

Other disproofs of relativity theory

Relativity theory has successfully acted as a basis for quantum electrodynamics, largely because it is self-consistent. There have been many criticisms of relativity theory, however, and generally leading scientific journals have been reluctant to publish them.

Browsing the Internet, however, quickly reveals a growing discontent with STR. For example:

[1] A paper posted on the Internet by Robert Sungenis gives convincing arguments against the most basic tenets of STR. He argues that

(a) the speed of light is not constant and independent of the position of the observer.

(b) time is absolute, no matter whether a clock slows in a different environment, such slowing being attributable to the different environment and the time taken to communicate between the moved clock and base of observations.

(c) Lorentz contraction is merely an observational phenomenon, and rods moving at high speeds will not, in fact, contract in absolute terms.

Sungenis concludes that STR is an "ad hoc invention", which led to the also erroneous GTR and also the Big Bang theory (Sungenis 2012).

[2] In a paper posted on the Internet, Francisco Muller (Muller 2013), discusses the electrodynamics of a pair of magnets, one fixed to a laboratory and the other moving. According to STR, the induced field calculated as the vector cross product v X B, where v is the velocity of the magnet, and B is the strength of its magnetic field, is always zero.

When the magnets have "transversal edges", however, these prevent induction whereas when they do not have these edges, induction occurs, disproving STR.

[3] A commentary posted on the Internet at math.ucr.edu/home/baez/physics/Relativity/SpeedOfLight is brief but to the point:

To state that the speed of light is independent of the velocity of the observer is very counterintuitive.
Given Einstein's starting point, it leads to him changing what is meant by space and time. There is no reason for making such a change. Therefore his starting point is wrong.

[4] Richard Calkins, author of *Relativity Revisited*, notes that the STR was based on the assumption that light was isotropic, having the same speed in all directions, and that light was subject to the same relativistic effects as the motions of physical objects.

In contrast, he argues that movements should be deemed relative to a stationary universe-wide reference frame of two lasers in which each point in that frame is 'at rest' and movements of any other reference frames can be defined relative to any of these stationary reference points.

Then because laser beams are a vector quantity, not the scalar quantity 'c' in the STR, they invalidate the STR assumption that electromagnetic phenomena such as the propagation of light are subject to the same relativistic laws as the fundamental quantities length, mass, and time.

He concludes that all light photons, once emitted, also continue to propagate in the direction of emission, and are thus a vector quantity, and that treating light as a vector quantity invalidates the prediction of STR.

Corrected theories of relativity

In a paper in *The General Science Journal*, Harry Ricker III argues that the logic behind the Lorentz-Fitzgerald contraction is invalid and that, rather than the length of a moving rod changing, what changes is the scale of measurement in the moving system relative to a stationary one (Ricker 2014). Ricker presents a new method which corrects this anomaly, concluding:

In this paper a new method was used that removes the difficulty by formulating the codomain length as a function of domain length and codomain time coordinates. However the solutions derived from the new method do not support the traditional interpretation of the theory of relativity that there is a physical contraction of space.

Buenker shows that the Fitzgerald-Lorentz contraction (FLC) is an observational artifice and that, locally and absolutely, no contraction in length actually occurs. He concludes that the FLC is a "myth" and proposes an alternative Lorentz transformation (ALT) to replace that used in the STR, this being adjusted to be consistent with the asymmetry observed in time-dilation experiments. He also proposes an amended relativity principle or ARP (Buenker 2013).

The Theory of Relativity

Lorentz pointed out that the condition c = constant only related two space-time vectors to the extent of a constant normalization factor between them.

Buenker includes a normalization factor N that is a function of the relative speed u of two reference frames, so that the Lorentz transformation is written as:

$$dt = N(dt' + udx'/c^2)/F$$

where, as before, $F = \sqrt{1 - u^2/c^2}$, and with only relative speed relating the frames N = 1. He then assumes that N can also be a function of the object's speed v', not just the relative speed u, allowing values of N other than unity.

Adopting the "principle of the simultaneity of events", we then have:

$$dt = N(dt' + udx'/c^2)/F = dt'$$

so that the normalization constant N is given by

$$dt/dt' = NT/F$$

where $T = (1 + udx'/c^2dt') = (1 + uv'/c^2)$
and N = F/T for simultaneity.

Then, including this normalization constant the alternative space-time transformation is obtained as:

$$dx = (dx' + udt')/T = N(dx' + udt')/F$$
$$dy = Ndy'; \quad dz = Ndz'; \quad dt = dt'$$

When N=1 the equations of the Lorentz transformation given in the second section of this chapter are obtained. Thus, according to Buenker (2007), they are "consistent with all known measurements that are otherwise claimed as proof of the validity of Einstein's original theory. But the ALT also leads to the conclusion that events occur simultaneously for all observers, regardless of their state of motion."

Indeed, in later work, not only does Beunker dispute the length contraction predicted by STR, but he argues that, according to the principle of constancy of the speed of light inherent in the STR, that if the clock on the relatively moving system runs slower, then there should also be isotropic length increase in the moving system (Buenker 2011).

The large curvature correction in FEM

Matrix structural analysis was developed by aeronautical engineers in Germany during the 1940s and extended to develop the finite element method (FEM) in the USA during the 1950s.

In analysing structures loaded to such an extent that they undergo large displacements calculation of the extensional strains is adjusted to include a term arising from movement transverse to the extensional strain.

In 1979, the first author introduced a calculation to take account of large curvatures in parts of structures subject to significant bending (Mohr and Milner 1979). Details of this large curvature correction (LCC) are given in Appendix C, but summary details are given here.

If curvatures are calculated according to a Cartesian frame of reference, a beam's curvature is given by the well-known formula:

$$C = d^2v/dx^2/F^3 \text{ where } F = [1 + (dv/dx)^2]^{1/2}$$

Because large displacements significantly change the geometry of the structure being analysed, the problem becomes nonlinear, and a stepwise solution using 'residual loads' is required.

To calculate these large displacements a 'virtual' curvature increment is calculated as

$$\begin{aligned}V(C) &= V(d^2v/dx^2)F^{-3} + (d^2v/dx^2)V(F^{-3}) \\ &= V(d^2v/dx^2)F^{-3} - 3(d^2v/dx^2)(dv/dx)F^{-5}V(dv/dx) \\ &= V(d^2v/dx^2)[1 - 3(dv/dx)^2/F^2]/F^3\end{aligned}$$

here $V(\ )$ denoting variation in $(\ )$ and using reciprocity to exchange incrementation in dv/dx with incrementation in d^2v/dx^2 in the second term.

The Theory of Relativity

Alternatively, the beam's curvature can be calculated with respect to its curvilinear coordinate s, which follows the curved shape of the beam:

$$C = (d^2v/ds^2)/F \text{ where } F = [1 - (dv/ds)^2]^{1/2}$$

Then a virtual increment in this is given by:

$$\begin{aligned}V(C) &= V(d^2v/ds^2)F^{-1} + (d^2v/ds^2)V(F^{-1}) \\ &= V(d^2v/ds^2)F^{-1} + (d^2v/ds^2)(dv/ds)F^{-3}V(dv/ds) \\ &= [V(d^2v/ds^2)/F][1 + (dv/ds)^2/F^2]\end{aligned}$$

and again reciprocity is used to obtain the final result.

That one can choose either an orthogonal Cartesian frame of reference, or a curvilinear frame of reference that follows the curvature of the deflected beam is reminiscent of the GTR in which curved space-time is proposed.

For practical purposes, as shown in Table C.1, the Cartesian LCC gives as good, if not better, results than the curvilinear LCC. To confirm the LCC to high accuracy, however, an extra 'degree of freedom' at each 'node' in the FEM model was used, and the curvilinear LCC gives better results when this is used, as shown in Table C.2, the results of Table C.3 confirming the LCC beyond doubt.

Conclusions

The contraction in length predicted by the theory of relativity has never been proven. Indeed, many 'disproofs' of this contraction have been produced, a couple of them being noted in the foregoing chapter. Indeed, observer-based length correction is merely a sort of parallax correction. In fact, objects do not contract in absolute terms as a result of their speed, however great.

Thus predictions of infinitesimal size and infinite mass as a particle gets close to the speed of light are absurd. These same predictions, however, are the basis of the assumption of the 'singularity' of the Big Bang.

Chapter 6

In the strange world of particle physics collisions and transformations of high-speed particles that may result in large changes in mass occur, of course, when relatively massless particles combine with much more massive ones. These changes are not, however, a result of relativistic phenomena as often claimed.

Indeed, recent experiments at CERN have shown that neutrinos can travel faster than the speed of light, contradicting perhaps the most fundamental pillar of relativity theory. A further contradiction has been observation of photons from distant stars with two speeds.

Today there are many who dispute the theory of relativity, and some of the 'disproofs' of it have been discussed in the foregoing chapter.

Of particular note is Buenker's alternative theory which makes time an absolute and does not predict the 'disproved' length contraction of the STR.

Indeed, if we accept time as absolute, as we should, then the need to adjust the rates of earthbound and satellite atomic clocks of the GPS system must be attributed not to time dilation, but to other field effects such as gravity, just as the redshift attributed to expansion of the universe is more sensibly attributed to other causes, as discussed in Chapter 9.

Similarly, as it has been clearly shown that relativistic length contraction does not occur, and that time should be deemed absolute, the 'singularity' or Big Bang model of the evolution of the universe is invalid. Indeed, belief that a virtually infinite universe can have originated from an infinitesimal and infinitely dense singularity is ridiculous in the extreme, and the Big Bang theory and alternative theories are discussed in later chapters.

As for Einstein's famous prediction of deflection of light by the sun's gravity, a simple approximate calculation of this is given in Appendix B, illustrating how this small effect should indeed occur.

In conclusion, widespread belief in the theory of relativity and all its predictions is comparable to religious beliefs. The mathematical language differs from the religious propaganda and hype that has fooled most people for millennia (Mohr and Fear 2014), but is equally effective in duping both those familiar with it, and the general public. In other words, propose only a

The Theory of Relativity

single equation, however important, and history may not take much notice. Propose a 'theory', such as that of animal and plant evolution, however, and that might be noticed.

That the theory of evolution is essentially correct, however, is in little doubt, but the theory of relativity seems likely to require considerable adjustment if it is to survive even in part.

In particular, time must be considered absolute, and for normal practical purposes an orthogonal frame of reference should suffice. After all, it has been confirmed that the universe is a flat 3-D Euclidean space.

Even when calculations are more accurate using a curvilinear coordinate system, as in the case of Mohr's large curvature correction when four or more freedoms/node are used, it would be ridiculous and impractical to then propose that our world view should always be in some 'bent' form. In other words, beams bend, and light can be diverted, but our frame of reference when an architect plans a building, or when we consider space, needs to remain a single orthogonal frame of reference at the point of observation.

Chapter 7

The Steady-State Theory of the Universe

> The main efforts of investigators have been in papering over contradictions in the big bang theory, to build up an idea which has become ever more complex and cumbersome.
> I have little hesitation in saying that as a result a sickly pall now hangs over the big bang theory.
>
> Fred Hoyle, *The Intelligent Universe.*

Introduction

In the 1920s many scientists resisted the introduction of general relativity into cosmology. Among them were William MacMillan, Svante Arrhenius, Emil Wiechert, Walther Nernst, and Robert Millikan, and they were opposed to predictions of an eventual collapse of the universe, believing that the universe was infinite in both space and time. Thus, Chicago astronomer MacMillan denied

> *"that the universe as a whole has ever been or ever will be essentially different from what it is today,"* saying *"the distribution of matter throughout space is uniform in the sense that . . . this portion*

The Steady-State Theory of the Universe

of the physical universe which comes under our observation is not essentially peculiar."

In 1929, Edwin Hubble reported observations that distant galaxies had displacements in the wavelengths of their light toward the longer red end of the spectrum ('redshifts').

When galactic redshifts were attributed to expansion of the universe, Nernst and his allies suggested alternative explanations for redshift based on a static universe. Nernst was also first to hypothesize a zero-point cosmic radiation.

The steady-state theory

In 1928, Sir James Jeans proposed his continuous-creation theory in which matter is continuously created throughout the universe, having before then studied spiral nebulae, the source of stellar energy, binary and multiple star systems, and giant and dwarf stars.

Jeans disputed the nebular hypothesis of Laplace that the sun and planets condensed from a single gaseous cloud, instead supporting the theory first suggested by American geologist Thomas Chamberlin that a star passing close to the sun had attracted stellar debris that condensed to form the planets.

In 1948, Hermann Bondi, Thomas Gold, and Fred Hoyle proposed the steady-state theory of the universe, proposing

(1) that the universe is, in terms of the large scale distribution of stars and planets, much the same at any point.
(2) that the universe stays much the same through time.
(3) that the mean density of matter in the expanding universe is maintained by continuous creation of matter.

They had therefore adopted Jean's 1928 continuous-creation theory to deal with widespread belief that the universe was expanding that observation of galactic redshifts and engendered.

(Note that alternative explanations of redshift are discussed in Chapter 9.)

Chapter 7

Problems for the steady-state theory

Hoyle et al.'s steady-state theory had dealt with the attribution of redshifts to expansion of the universe by adopting Jeans' continuous-creation of matter hypothesis, proposing the 'C-field' as a "matter-creation field".

Proponents of the earlier Big Bang theory attacked hypothesis of the C-field as requiring unknown and perhaps non-existent physical mechanisms.

Another aspect of the redshift question is Olbers' paradox. Named after German astronomer Heinrich Olbers, this paradox is that the night sky is dark whereas in a static infinite universe it should be bright because the total light from nearer regions of the sky would be the same as that from more distant regions.

Proponents of the Big Bang theory argue that Olbers's paradox occurs because distant galaxies are moving away from us faster than nearer ones, in line with Hubble's law, so that much of the visible light from distant galaxies is redshifted to the invisible infrared and microwave regions.

An alternative explanation is that the universe is not transparent, and light from distant galaxies is absorbed by cosmic dust and this explanation is discussed further in Chapter 9.

A further problem for the steady-state theory came with Sir Martin Ryle's counts of extragalactic radio sources during the 1950s. Ryle found more radio galaxies at large distances, suggesting that the spatial and time distribution of matter was not uniform, as held by the steady-state theory. In addition, the Big Bang model predicts a deficit of faint sources whereas observations showed a surplus of faint sources. The steady-state model predicts an even greater deficit and cannot be adjusted to improve its prediction. The Big Bang model, however, involves different periods, for example that of inflation, allowing adjustment of some of its predictions.

Another problem for the steady-state theory was that the Big Bang theory provided a 'more direct' explanation for the relative abundance of light elements in the cosmos.

In Big Bang nucleosynthesis, only isotopes of hydrogen, helium, and a trace of lithium are produced, consistent with the proportions of elements in the universe today. In the original steady-state model all of the heavy elements were produced in stars by burning hydrogen into helium, helium nuclei (alpha particles) then combining to form heavier nuclei such as carbon (3 alpha particles) and oxygen (four alpha particles). The original steady-state model could not explain the relative abundance of helium in the universe, and in a paper in the journal *Nature* in 1964, Hoyle and Tayler proposed that most of the helium in the universe was produced by explosions in supermassive objects, not in ordinary stars.

Opinion remained fairly evenly divided between the Big Bang and steady-state models, however, until discovery of the cosmic microwave background radiation (CMB), this being deemed by supporters of the Big Bang theory to be a remnant of much stronger radiation dating from soon after the Big Bang. Moreover, detailed studies in the 1990s showed that the CMB was closer to that of a blackbody than any other source in nature.

Proponents of the steady-state theory proposed that particles of the interstellar medium, including 'whiskers' of iron, carbon, and bacteria, could scatter radio waves in such a manner as to produce an isotropic CMB (Hoyle 1983). In addition, steady-state theory did not have the 'horizon problem' of the Big Bang because it allowed an infinite amount of time for the 'thermalizing' of background radiation.

Quasars

That the great majority of quasars, or quasi-stellar radio sources, are very distant contravenes the isotropy assumption of the steady-state theory.

Quasars are millions of times more massive than the sun but occupy a region around the size of our solar system. It is widely believed that quasar-like regions might be responsible for the explosions that sometimes occur at the centres of galaxies. These explosions involve high temperatures and densities in the same way as proposed in the early stages of the Big Bang (Hoyle 1963), an observation that may have given encouragement to the 'little bangs' proposed in the QSSC model discussed in the following section.

Chapter 7

The quasi-steady state cosmology (QSSC)

Throughout history, cyclical theories of the universe have been considered. In 1923, Alexander Friedmann wrote:

The universe contracts into a point (into nothing) and then increases its radius from the point up to a certain value, then again diminishes its radius of curvature, transforms itself into a point, etc.

For this cyclical model of the universe he calculated a period of about 10^{10} years, based on the assumption of a zero cosmological constant and the mass of the universe being 5×10^{21} sun masses.

The QSSC was an attempt by Hoyle, Burbidge, and Narlikar to explain the surplus of faint radio sources and the evolution of the CMB by proposing that the growth rate of the universe varied sinusoidally over time but with a diminishing amplitude of variation The period of these oscillations were taken to be about three times the age of the Big Bang universe (Hoyle et al. 1993).

In this model, there was a 'creation field' with an initially negative energy density that scales like radiation and becomes dominant at high redshifts. The 'bursts' of creation that this theory suggests are sometimes referred to as *minibangs, mini-creation events,* or *little bangs.*

The cyclic nature of the QSSC provides an explanation for redshift, namely that the universe is currently in an expansion phase.

After the observation of accelerating expansion of the universe, further modifications of the model were made.

Again the CMB is held to be caused by carbon and iron whiskers in the interstellar medium, but opponents of SS theory argue that

(a) radiation as smooth as the CMB could not have come from point sources.
(b) the CMB does not show features such as polarization normally associated with scattering.
(c) superposition of radiation from dust clumps at difference temperatures and redshifts could not be so close to that of an ideal black body as is the CMB radiation.

Arguments for and against the QSSC model continue to appear, though proponents of the QSSC are greatly outnumbered by those of the Big Bang.

Like the QSSC, *chaotic inflation theory* proposes an infinite universe and continuous creation of matter, rather than most matter being created soon after the Big Bang. Unlike SS and QSSC, however, the chaotic inflation theory proposes that inflation operates only outside the observable universe, making this theory unprovable.

New galaxies

Some support perhaps for the ongoing C-fields proposed by steady-state theory, NASA's Galaxy Evolution Explorer found what appeared to be three dozen massive galaxies in Earth's corner of the universe that may be as little as 0.1 to 1 billion years old, our Milky Way being 10 billion years old. The new galaxies emit about ten times more UV than the Milky Way, indicative of their recency.

According to the Big Bang theory, most of the galaxies in the universe, if not all of them, should have been formed very early in time. That many very recent "baby" galaxies have appeared in just one part of our galaxy seems to contradict the Big Bang hypothesis.

Conclusions

There have always been creationist theories of the universe such as that of the Big Bang. Indeed, Hoyle argued against the latter, claiming that it was creationist dogma, coining the term Big Bang in the process, and thereby doing much to ensure its eventual dominance.

Indeed, as Hawking noted,

The Catholic Church, on the other hand, seized on the big bang model and in 1951 officially pronounced it to be in accordance with the Bible (Hawking 1989).

Chapter 7

Proposed as an alternative, the steady-state model and its modification, the QSSC model, are now only supported by a minority because steady-state models have greater difficulty in explaining:

(a) redshift of light from distant stars, though the QSSC model does explain redshift as the result of the universe currently being in an expansion phase.
(b) the very uniform and 'blackbody like' cosmic microwave background radiation.
(c) the relative abundance of the lighter elements such as helium and lithium.

Findings that support steady-state theories include

(a) that quasar-like objects might be responsible for the 'little bangs' that occur at the centre of galaxies.
(b) the many young galaxies recently found in a corner of our galaxy disprove the Big Bang concept of almost instantaneous creation of the universe.

A good deal more thought needs to be given to these matters and the Big Bang, redshift, the CMB, and new theories of the universe are discussed in following chapters.

Chapter 8

The Big Bang Theory

> *The mass starts into a million suns;*
> *Earths around each sun with quick explosions burst,*
> *And second planets issue from the first.*
>
> Erasmus Darwin, *The Botanic Gardens*.

Introduction

In 1922, Russian meteorologist and mathematician Aleksandr Friedmann argued that Einstein's general theory of relativity required a changing universe, rather than a static one. Assuming a closed universe, he hypothesized a Big Bang followed by expansion, then contraction with an eventual big crunch.

Other proponents of the Big Bang model assume an open universe that expands indefinitely, or a flat universe in which expansion gradually slows to a halt.

In 1927, Belgian scientist Georges Lemaitre also used the general theory of relativity to model an expanding universe.

In the 1920s, Edwin Hubble began studying bright galaxies looking for Cepheid variables, supergiant stars that pulsate regularly and are hundreds of times brighter than the sun, a star's true brightness being the amount of light it emits regardless of its distance from Earth.

Knowing the true brightness of Cepheids, Hubble was able to calculate how far they were from Earth. He then found that their spectra were shifted towards the higher red wavelengths, the amount of this 'redshift' being proportional to their distance from Earth.

Attribution of redshift to recession velocities (the Doppler effect) led in 1929 to Hubble's law, which is detailed in the last section of Chapter 3, and according to which, the greater the distance of a star from Earth, the greater its apparent velocity, a result which encouraged development of the Big Bang theory.

Estimates of the Hubble constant have varied greatly and obtaining an accurate estimate of this was the key aim of the Hubble Space Telescope (HST) project, resulting in the current estimate of the age of the universe being 13.7 billion years (Sparrow 2006).

Development of the Big Bang model

In the 1940s, theoretical physicist George Gamow began working out details of Friedmann's solutions to Einstein's theory, directing his research students Ralph Alpher and Robert Herman to study likely nucleosynthesis reactions in the early universe.

They expanded on Gamow's idea that the universe expanded from a primordial state of matter consisting of protons, neutrons, and electrons in a sea of radiation. They theorized that the universe was very hot at the time of the Big Bang, because elements heavier than hydrogen can only be formed at a high temperature.

Alpher and Hermann also predicted that radiation from the big bang should still exist and cosmic background radiation roughly corresponding to the temperature they predicted was detected in the 1960s (Singh 2005).

The Big Bang model

The initial Big Bang took only the tiniest fraction of a second and temperatures reached trillions of degrees, allowing matter and energy to switch back and forth.

In the first one-billionth of a second after the Big Bang (AB) the first heavy particles, quarks were formed. In the next fraction of a second the first lighter particles, leptons were formed.

Of the six types of quarks, only the two lighter 'up' and down' quarks are stable, and as temperatures fell after about one microsecond these joined together to form protons and neutrons in a ratio of about seven to one.

Similarly, only two of the six types of lepton, the electron and the almost massless electron neutrino, are stable at low temperatures.

After one second temperatures were too low for further spontaneous creation of matter from energy and the universe was a seething mass of quarks, protons, neutrons, electrons, neutrinos and photons. Within 100 seconds AB, protons, and neutrons began to join to form simple nuclei, but most of the proton excess remained alone as nuclei of the simplest element, hydrogen.

After about 300,000 years AB, the temperature had decreased to a few thousand °C and electrons combined with nuclei to form atoms and the universe became clear enough for the transmission of light over long distances (May et al. 2006, Sparrow 2006).

Inflation

Big Bang theory accounts for the high proportions of light elements such as hydrogen and helium, for the expansion of the universe which is apparent as one explanation for redshift, and the cosmic background radiation.

The theory has also encountered several problems, and been adjusted several times to deal with these.

The most important of these was that, prior to 'clarification' of the universe circa 300,000 AB, the pressure of trapped photons should have prevented matter amalgamating to form the first atoms. Then clarification of the universe would have taken place much later, by which time matter would have been too evenly spread out to allow the inhomogeneity found in today's universe to develop.

To explain this inhomogeneity, in 1981, physicist Alan Guth proposed that in the first 10^{-32} second the universe expanded exponentially to 10^{50} times its original size, this period of 'inflation' allowing the universe to rapidly fill most of the space it presently occupies, but with small inhomogeneities that grew over time, explaining why matter in the universe is not evenly distributed.

Chapter 8

The discovery by the Cosmic Background Explorer (COBE) in 1992 of small temperature and density variations in the CMB was interpreted by proponents of BB theory as evidence of inflation after the Big Bang.

After inflation, the universe contained a quark-gluon plasma, as well as all other fundamental particles. Very high temperatures caused random motions of particles at relativistic speeds, and particle-antiparticle pairs of all kinds were continuously created and destroyed by collisions and a slight excess of matter over antimatter resulted.

The Lambda-CDM model

The Lambda-Cold Dark Matter model of the Big Bang is often referred to as the 'standard model' of Big Bang cosmology and this is extended to incorporate inflation.

It incorporates the cosmological constant, denoted by the Greek letter lambda, which is associated with 'dark energy' in space which is postulated as the cause of the currently accelerating expansion of the universe.

Until about five billion years ago expansion of the universe was slowing down because of the gravitational effects, but since that time expansion has been accelerating, and it was postulated that this was because the universe had expanded to the point at which the repulsive forces of dark energy began to exceed the attractive forces of gravity.

The four pillars of BB theory

The four key pillars of the Big Bang theory are the following:

[1] Redshift in light from distance parts of the universe and attribution of this in Hubble's law to expansion of the universe. Misleadingly termed 'metric expansion', this increases the distance between objects that are not under shared gravitational influence, but does not increase the size of objects such as galaxies in space.

The Big Bang Theory

[2] The existence of the cosmic microwave background radiation, supposed to be from the 'surface of last scattering' near the end of inflation, and its close resemblance to that from almost perfect black bodies.
[3] The relative abundances of hydrogen, helium and lithium in the universe supposed to have arisen in the earliest nucleosynthesis reactions after the Big Bang.
[4] The large-scale structure of the universe which is spatially almost flat and in which on a larger than galactic scale is relatively homogeneous.

The general theory of relativity is an important part of the Big Bang theory, having been argued by Friedmann to require a changing universe, not a static one.

Lemaitre and others since have also used the general theory of relativity to model an expanding universe that began with the Big Bang.

Problems with the Big Bang

The Big Bang theory explains observed phenomena such as expansion of the universe, but leaves many things unexplained, such as the following:

[1] The light-horizon problem. The usual example of this is to consider CMB coming from two opposite sides of the sky. These photons, travelling at the speed of light since their emission by plasma, have only had time to reach the Earth now and not time to have "communicated" so how do they "know" to have the same temperature to a precision approaching one part in 100,000? In other words, what accounts for the high degree of angular isotropy of the cosmic microwave background?
[2] The flatness problem. The universe has been found to be almost flat in space-time, an unlikely outcome from a Big Bang origin.

[3] The monopole problem. Particle physics uses the grand unified theories to propose that at the very high temperatures immediately after the Big Bang the strong, weak, and electromagnetic forces were unified into a single force. Such a force should have produced massive particles called *magnetic monopoles*. If so, some of these should still be evident but no such particle has been found.

The theory of inflation solves these problems, in principle at least. Thus, in a period called the *inflationary epoch*, a 'phase transition' in which the nuclear force breaks away from the unified force is thought to have occurred 10^{-35} seconds after the Big Bang created the universe. This transition filled the universe with very dense 'vacuum energy' which made gravitation become repulsive for about 10^{-32} seconds, during which period the universe 'inflated' by a factor of about 10^{50} from a size of only a few cm before inflation to produce all that we can see today.

Inflation solves the three problems above as follows:

[1] Before the great expansion of inflation, widely separated regions we see now were close together and had the uniformity evidenced in the CMB.
[2] Inflationary expansion greatly reduced any initial curvature in space-time.
[3] Inflationary expansion dispersed any magnetic monopoles so greatly that they are virtually impossible to detect.

Inflation is also believed to have been capable of producing the small density fluctuations that later allowed matter to clump together to form galaxies.

Conclusion

The Big Bang model of cosmology holds sway at present, though there are still many adherents of modified steady-state models. Other alternative models have also been proposed, and some of these are discussed in Chapter 12.

The Big Bang Theory

There also many who dispute that the universe is really expanding, and this issue is discussed in Chapter 9, alternative explanations for the CMB other than the Big Bang being discussed in Chapter 11.

Chapter 9

Is the Universe Really Expanding?

> As astronomical red shifts vary with wave length (dispersion) they cannot be due to recession of a single object. Therefore Hubble redshifts cannot be due to expansion of the universe.
>
> R.I.P. Expanding Universe (b. 1930, d. 2012), David Noel.

Photons and light

Photons are stable elementary particles which carry electromagnetic force. They behave as waves when refracted by a lens or when there is interference between light waves. They also behave as particles and are only emitted and absorbed at certain quantized energy levels.

Photons are stable, have no charge, and are generally assumed to be massless, though some calculations show that they may have a minute mass. Photons inside superconductors develop nonzero effective rest mass, and under these circumstances, electromagnetic forces become short range.

The wavelength of visible light varies from 400 (blue/violet) to 700 (red) nanometers. Wavelengths just outside this range are respectively called ultraviolet and infrared.

About half the energy of sunlight at the earth's surface is visible electromagnetic waves, about 3% is ultraviolet, and the rest is infrared.

Is the Universe Really Expanding?

X rays have wavelengths ranging from several thousandths of a nanometer to several nanometers, gamma rays have even shorter wavelengths. Microwaves have wavelengths ranging from 0.0004 metres to about 0.1 metres, and radio waves have wavelengths ranging from about 0.1 metres to over three hundred metres.

Hubble's law

Until the early 1920s, it was believed that our galaxy was at the centre of an otherwise empty universe. As telescopes became more powerful and spectrometry improved, however, many other galaxies were discovered.

The discovery that light from these galaxies was weakened more than expected by being shifted in wavelength towards the red end of the visible spectrum was made by V. M. Slipher (Hoyle 1983).

These 'redshifts' were attributed to the Doppler effect, which is given by the formula

$$f = f_0 c/(c + v_s)$$

where f_0 is the frequency actually emitted by the source, c is the velocity of light, v_s is the velocity of the source relative to the observer, and f is the frequency detected by the observer.

Thus it was assumed that redshifted galaxies were moving away from us, apparently at great speed.

The work of Edwin Hubble and his assistant Milton Humason extended that of Slipher. They found that the most distant galaxies were retreating more rapidly, leading to Hubble's law:

$$v_s/D = H$$

where D is the distance of the object, and H is the Hubble constant, and an example calculation using this law is given in Chapter 3.

Hubble initially estimated the Hubble constant as being 500 km/sec/megaparsec, where 1 parsec = 3.26 light years, and 1 megaparsec = 3.0856×10^{22} metres.

This has since been radically revised, one estimate in the early 1990s being only 50 km/s/Mpc, another at that time being 100 km/s/Mpc. In 2011 NASA estimated the Hubble constant to be 73.8 +/-2.4 km/s/Mpc.

It was Hubble's law that led to belief that the universe was expanding, and thence to the Big Bang theory of the universe, even though it has been said that Hubble himself believed that the recession velocity associated with redshift was an 'apparent' effect that did not actually occur, the actual cause of redshift being unknown.

Redshift, as a result of expansion of the universe, is called cosmological redshift, but there are several other processes that can cause redshift.

Accelerating expansion of the universe

Astronomers have been puzzling over the expansion rate of the universe and its density for decades. If its density is large enough, the universe will eventually begin to shrink and ultimately collapse. If the density is less than a certain critical density, however, the universe will continue to expand indefinitely. The ratio of the actual density to the critical density is called Omega.

The inflationary Big Bang model predicts that Omega is one and that expansion of the universe should slow down.

Observations of type 1a supernovas, however, suggest that expansion of the universe is accelerating, one study of almost sixty supernovas showing that nearly all type 1a supernovas are at least 15% farther away than the standard model of the universe predicts

Such findings suggesting accelerating expansion of the universe led to the hypothesis that a new force, dark energy, which was now stronger than and opposed to gravity, was causing expansion of the universe to accelerate.

Relativistic redshift

General relativity predicts that as radiation travels through space, its wavelength increases to keep up with the expansion of the universe. This mechanism involves a progressive stretching occurring throughout light's journey from a distant source, rather then being caused by relative motion between the source and the observer. Therefore, light from more distant galaxies that has been travelling longer has greater redshift.

Relativistic redshift can be used as a measure of the relative size of the universe when the light was emitted, radiation of redshift Z having been emitted when the universe was the fraction $1/(1 + Z)$ of its present size width (Ridpath 2001).

Gravitational redshift

Another mechanism for redshift is the gravitational redshift (GRS) which shifts radiation moving out of a gravitational field. It was predicted by Albert Einstein's general relativity theory, according to which periodic processes are slowed down in a gravitational field. GRS is noticeable in the spectra of compact massive stars, such as white dwarfs.

Though the large redshifts observed in quasars are generally believed to be cosmological, some scientists believe that they may be a result of GRS or another unknown mechanism.

There are many other physical processes that can cause a shift in the wavelength of electromagnetic radiation, including scattering and optical effects. The resulting changes are distinguishable from true redshift and are not generally referred to as such.

The tired-light effect

The tired-light effect was proposed by Fritz Zwicky in 1929 as an alternative explanation for cosmological redshift. This effect was one of weak gravitational interactions of photons with stars and other material that reduced their momentum, resulting in redshift.

Indeed, Hubble himself favoured the tired-light theory, admitting that it would involve "some hitherto unrecognized principle in physics."

In the 1970s, Paul LaViolette developed a unified field theory called subquantum kinetics to explain the formation of subatomic particles.

In 1978, he found that his theory suggested that photons must lose energy as they travelled through space and thus exhibit a tired-light effect.

Studying the literature, he found that previous studies compared the expanding universe hypothesis and its velocity-redshift interpretation against only one set of data, eliminating any contradiction of expansion predictions with assumptions of hidden mass in the universe, or by introducing 'evolutionary corrections' such as assumptions about how the luminosity or size of a galaxy might vary over time.

He set out, therefore, to subject the tired-light theory and the Big Bang expanding universe model to four cosmological tests. His results showed that the stationary tired-light model consistently fitted the data better than the expanding universe model.

After some delay because of 'establishment objections', his paper on the work was published in the August 1986 issue of *Astronomy,* and LaViolette received many letters from notable astronomers and astrophysicists commending his work.

GRS and not an expanding universe

In a paper posted in the Internet, David Noel argues that gravitational effects are largely responsible for the redshifts attributed to recession velocities and the Doppler effect (Noel 2012). He notes that measured redshifts from distant galactic objects show dispersion, that is, they vary with wavelength, whereas the Doppler effect does not involve dispersion so that wavelength does not appear in the formula for the Doppler effect given earlier in this chapter.

Indeed, Noel had NASA astrophysicist Hans Krimm verify this (Noel 2012):

> Noel: *Have any checks been made of the spectra of red shifts from very distant galactic objects, to see if the red shift varies with wavelength? I am aware that current Doppler theory says there should be no dispersion, I would like to know if this theory is borne out by actual measurements, preferably of distant objects for which wide spectra are available (radio to x-ray).*

Is the Universe Really Expanding?

> Krimm: *There have been redshift measurements made for many distant galaxies across a wide spectral band (though mostly in the optical to infrared) and a consistent redshift with wavelength has been found. In fact, this is one of the ways in which astronomers confirm that the spectral shifts they are seeing are redshifts—the same delta-wavelength is measured for multiple spectral lines.*

Noel concludes that the GRS explanation of redshift does not require the assumption of the Doppler effect or an expanding universe, but that any expanding universe theory must take account of GRS.

Noel also offers an alternative explanation to that of the Big Bang for cosmic microwave radiation, and this is discussed in Chapter 11. His 'Placid Model' of the universe is discussed in Chapter 12.

Olbers' paradox

As noted in Chapter 7, this is that the night sky is dark, whereas in a static infinite universe, it should be bright because the total light from nearer regions of the sky would be the same as that from more distant regions.

Proponents of the Big Bang theory argue that Olbers's paradox occurs because distant galaxies are moving away from us faster than nearer ones, in line with Hubble's law, so that much of the visible light from distant galaxies is redshifted to the invisible infrared and microwave regions.

An alternative explanation is that the universe is not transparent, and light from distant galaxies is absorbed by cosmic dust so that the distance from which light can reach the observer is bounded. Such absorption, however, should heat the cosmic dust and it should reradiate energy, albeit with some loss of intensity according to the second law of thermodynamics.

In fact, according to quantum mechanics, submicron cosmic dust should be able to redshift light to infrared and microwave regions of the electromagnetic spectrum without increasing in temperature. Therefore, Olbers's paradox can be explained by the absorption of light by submicron cosmic dust in a static infinite universe.

Indeed, cosmic dust is one of the most frequently cited alternative causes of redshift to that of expansion of the universe. A key objection is that light scattered by cosmic dust should give blurred images, whereas galaxy images are not blurred. The QED-induced light theory discussed in the following section overcomes this objection.

The QED induced light theory

The quantum electrodynamic (QED) theory of redshift considers a galaxy photon absorbed by a cosmic dust particle as confined within the dust particle geometry.

According to quantum mechanics (QM), photons of wavelength L_0 are created by supplying EM energy to a QM box with walls separated by $L_0/2$. For a spherical dust particle of diameter D, the QED photons are produced at wavelength $L_0 = 2Dn$, where n is the index of refraction.

For amorphous silicate n = 1.45, giving for dust particles with D = 0.25 microns a redshift of the Lyman-alpha line at 0.121 microns to a red line at 0.725 microns with redshift $Z = (L_0 - L)/L = 4.8$. Such a redshift would be interpreted by Hubble's Law as implying a recession velocity over 90% that of light when, in fact, the cause may have simply been cosmic dust.

QED redshift from supernovas

The QED theory also easily explains the time dilation of supernova light curves at high redshift, for example redshift =1 will make a supernova explosion that actually takes twenty days appear to take forty days. Redshift in the QED induced light theory is proportional to the number of dust particles in the light path, this in turn being proportional to the total mass emitted by explosion of a supernova.

Time dilation in observing supernova explosions is simply a result of cooling of the dust mass, larger dust masses taking longer to cool resulting in greater redshift.

Light from the interstellar medium

The infrared spectrum observed in the interstellar medium (ISM) is usually attributed to thermal emission from dust particles heated by UV photons. Visible light is also attributed to photoluminescence as a result of dust particles absorbing UV photons. These spectra are more likely produced by QED-induced redshift.

Visible colours in the ISM require dust particles less than 0.5 microns, while the IR spectra require larger dust particles found in molecular clouds. Thus, the vivid ISM colours observed can be produced by single UV photons at 0.15 microns being absorbed in silicate with D = 0.125 to 0.25 microns, resulting in blue to red light with wavelengths from 0.362 to 0.725 microns and redshift Z from 1.41 to 3.83. Therefore, redshift of light in the ISM could have been caused by cosmic dust, and not by expansion of the universe (Prevenslik 2010).

The Tolman test and QED-induced light theory

This was devised to test whether the universe was static or expanding. In a static universe both the apparent size and the apparent light intensity of an object decrease inversely with the square of its distance from the observer, so the brightness/unit area is independent of distance.

According to theory, if the universe is expanding the brightness of an object is reduced by the fourth power of (1 + Z), whereas the latest research finds that the exponent of (1 + Z) is 2.6 or 3.4, depending on the frequency band, the similarity of these reductions being claimed to imply expansion of the universe.

If the brightness B is reduced by absorption in cosmic dust particles this conclusion changes. A single QED interaction with a dust particle emits light at wavelength $L_0 = L(1 + Z)$. Then the brightness at the observer is $B_0 = hc/L_0 = hc/L(1 + Z) = B/(1 + Z)$, where h = Planck's constant, so that the brightness is reduced by (1 + Z). Then for multiple interactions, B decreases inversely with the product $(1 + Z1)(1 + Z2) \ldots (1 + ZN)$ where ZK is the redshift for interaction K. Thus QED theory can satisfy the Tolman test.

There is a limit, however, on the UV brightness of galaxies because when the surface density of hot bright stars such as supernovas increases, large amounts of dust are produced which absorb much of the UV.

One study compared the surface brightness values assuming either an expanding or static universe. It found that in the expanding case many galaxies would have been brighter than the limiting value by a factor of up to 6. In the static case, in contrast, no galaxy was brighter than the limiting value.

Angular size test

The angular sizes A of radio galaxies with a redshift up to Z = 2 are found to be proportional to the inverse of Z, a static Euclidean effect. The standard Big Bang model of the universe postulates that galaxies have grown in size to match the observed A proportional 1/Z. This would require galaxies at Z = 3.2 to have grown by an unlikely factor of 6.

Such a rate of growth would require between two and four mergers per galaxy during its lifetime, a highly improbable scenario. Furthermore, spiral galaxies, of which there are many, cannot grow by merging and still preserve their spiral shape.

The angular size test therefore casts further doubt on whether the universe is expanding.

Another study of galaxies with a wide range of redshifts (0.03 to 5.7) found that the angular size data fitted well with a static universe model, but poorly to the standard Big Bang model of the universe. When the redshift range was restricted to 0.03 to 3.5 the static model provided a good fit with the data, and when a very small amount of extinction was allowed for, the fit was nearly perfect (Hartnett 2014).

Conclusions

Attribution of redshift to the Doppler effect was the foundation stone of the expanding universe model.

In fact, redshift is found to vary with wavelength, indicative of dispersion, whereas the Doppler effect precludes dispersion.

Is the Universe Really Expanding?

Indeed, to propose that the farther away an object is the greater its recession velocity is clearly absurd. It is one thing to observe that objects in the universe emit light, quite another to propose that those farthest away from us, such as quasars, are travelling at almost the speed of light.

The most obvious and natural explanation of redshift increasing with distance, of course, is greater interaction of photons with cosmic dust. An objection to this proposal is that light scattered by cosmic dust should be blurred.

The QED-induced light theory overcomes this objection, showing that submicron dust particles can cause the observed galactic redshifts, and it also explains the IR and UV spectra in the interstellar medium.

Cosmic dust also provides an explanation for the time dilation of light from supernovas.

Gravitational effects may also contribute, at least in part, to redshift, this being the basis of tired-light theories.

Angular size observations cast much doubt on the expanding universe assumption but correlate well with a static universe.

Studies of very distant quasars show some behaviours, which contradict the standard Big Bang expanding model of the universe.

Thus evidence contradicting the Big Bang expanding universe hypothesis is mounting. In addition, one important advantage of the static universe model is that it does not require postulation of dark matter and dark energy.

Chapter 10

Is the Universe Infinite?

> *Recent measurements . . . have shown that the brightest spots* [in the CMB] *are about 1 degree across. Thus the universe was known to be flat within about 15% accuracy prior to WMAP results. WMAP has confirmed this result with high accuracy and precision. We now know (as of 2013) that the universe is flat with only a 0.4% margin of error. This suggests that the Universe is infinite.*
>
> Will the Universe expand forever?, NASA article
> www.map.gsfc.nasa.gov//universe/uni_shape.html

Introduction

Greek mathematicians such as Pythagoras felt numerical relationships were the key to understanding the world around them. Discovery of ratios such as pi, the ratio of the circumference of a circle to its diameter, was a surprise as the resulting numbers were not finite.

Circa 340 BC, Aristotle felt that the universe as we know it had emerged from an infinite and formless chaos and that the stars in the sky formed a massive sphere that encircled our world, this emergence of order from chaos being the basis of the first-cause argument for the existence of God.

Aristotle believed the Earth was spherical, estimating its diameter as about twice its actual size, and believed that the heavenly bodies moved in circular orbits around it.

Is the Universe Infinite?

In 1514, Copernicus proposed that the sun was the centre around which the Earth and its neighbouring planets revolved. Subsequently, Kepler proposed that these orbits were elliptical, and Newton's law of gravitation showed why such orbital motion occurred.

Newton felt that if there were a finite number of stars in a finite space that they should eventually coalesce, whereas if the universe was infinite with stars distributed relatively uniformly, such a collapse would not happen (Hawking 1989).

Averaged over large distances (larger than the largest galaxies) matter is spread uniformly throughout the universe, most of it being the lighter elements such as hydrogen, deuterium and helium. The average density is only 9.9×10^{-30} gm/cc, the relative fractions of chemical elements also being uniform. Only 4% of the mass-energy is ordinary matter, the rest being supposed to be dark matter or dark energy (*Wikipedia* 2014).

According to the 'cosmological principle', the universe is homogenous and isotropic and the same laws of physics and fundamental constants such as the gravitational constant G apply throughout.

In the general theory of relativity, the cosmological constant Λ represents a force to balance the attraction of gravitation and $\Lambda = 1$ for a steady-state universe. Some cyclic models of the universe assume $\Lambda = 0$.

The density parameter D for the universe involves the amount of matter in the universe and the Hubble constant, and $D = 1$ for the quasi-steady state cosmology (QSSC), thanks to its postulated creation-fields.

Some theories hold that if D is less than a certain critical density the universe will eventually collapse in a 'Big Crunch', others that if D is greater than this critical density it will expand forever, whilst cyclic models have varying values of D.

Is the universe expanding?

The preceding chapter discussed the question of whether the universe is really expanding, arguing that the redshift of radiation attributed to the Doppler effect is most probably caused by interstellar dust, that such redshifted light is not blurred being explained by the quantum mechanics of photon absorption and re-emission (Prevenslik 2010).

Chapter 10

Gravitational effects may also contribute to redshift, it being clear that if gravity can bend light, as in the general theory of relativity, then surely it can also change its wavelength under certain circumstances.

If redshift is not caused by the Doppler effect and recession velocities, then Hubble's law is invalidated as the general explanation of redshift so that the universe may not be expanding at all, and certainly it will not be expanding to the extent of distant galaxies receding away from us at nearly the speed of light.

If the universe was infinite, of course, it could not expand, but there would still be plenty of scope for movement, ranging from planetary orbits to the massive explosions of supernovas.

How big is the universe?

In a 2001 interview conducted at the European Space Agency's European Space Research and Technology Centre (ESTEC) in the Netherlands, Joseph Silk was asked the question: "Is the universe finite or infinite". He replied, *'We do not know whether the universe is finite or not.'*

In reply to the question how big the universe might be if it were finite, he replied, *'It could be as large as 100 times the horizon* [the distance light has travelled since the Big Bang, 13.7 billion years ago]. *That means that the universe would be as much as 100 thousand million parsecs, about 300 thousand million light years, if we could measure the topology.'*

The observable (from Earth) universe has a radius of about 46 billion light years and contains more than 100 billion galaxies, galaxies typically being 30,000 light years in diameter and 3 million light years apart. Galaxies contain between 10 million and 1 trillion stars, and in 2010 astronomers estimated that there were 300 sextillion (3×10^{23}) stars in the observable universe.

Thus, according to *Wikipedia* (2014): "The size of the universe is unknown; it may be infinite."

Is the universe infinite?

According to the general theory of relativity, the density of the universe determines its geometry. If the density exceeds the critical density, space is closed and positively curved like the surface of a sphere so that photons will follow curved paths. If the density is less than the critical density then space is infinite and negatively curved like a saddle. If the density equals the critical density, however, the universe is 'flat' and infinite.

The simplest version of the 'Big Bang + inflation' theory predicts that the universe's critical density is very close to the critical value, and thus that the universe is flat.

Fluctuations or 'spots' in the cosmic microwave background radiation (CMB) are believed to indicate the shape of the universe. If these are about one degree across the universe is flat, if less it is open, if greater it is closed.

On the largest scales, however, these variations in the CMB almost disappear, and some scientists believe this is because the universe is limited in size. They proposed various models for the shape of the universe, but having tested some of these models, it was reported of one scientist that "he suspects that the work will probably turn up nothing, meaning that the universe is either very large or infinite (Muir 2003).

Up to 2001, measurements by several ground-based and balloon experiments showed that these spots in the CMB were indeed 1 degree across within about 15% accuracy. The Wilkinson Microwave Anisotropy Probe (WMAP) confirmed this finding to within a 0.4% margin for error, suggesting that the universe is flat and thus infinite (NASA 2014).

Analysis of the results of more than three years of observations in the Baryon Oscillation Spectroscopic Survey (BOSS) by one of the world's widest-angle telescopes in New Mexico yielded the most accurate map yet of the universe showing 1.2 million galaxies with an accuracy of 99%. Presenting their findings at the 223rd meeting of the American Astronomical Society, the BOSS researchers said that their findings indicated that the universe was "extraordinarily flat" and was probably infinite in both space and time (Anthony 2014).

Chapter 10

Conclusion

There is growing evidence that the universe is infinite, or at least that it appears to be so. This is perhaps easier to imagine than a 'closed universe that is perhaps spherical or saddle-shaped in space-time.

The present authors, however, prefer to consider that the universe may be infinite in both space and time for 'all practical purposes'. They also strongly feel that the universe is not expanding according to Hubble's law and that the redshift attributed to the Doppler effect is largely caused by cosmic dust.

Here, 'practical purposes' are those concerning observations from our solar system whose sun is predicted by some scientists to have a limited lifetime of only another 4.5 billion years

Chapter 11
The Cosmic Microwave Background

> *I believe the myriad of fine particles that exist within the galaxies, and probably in the spaces between them, are prime candidates for making this transformation [to form the cosmic microwave background].*
>
> Fred Hoyle, *The Intelligent Universe*.

Cosmic background radiation

Cosmic background radiation is electromagnetic radiation in the sky with no discernable single source. It includes infrared, X-ray, and microwave components, the latter being discussed in the following section.

Redshifted light from nuclear fusion inside stars is the main source of cosmic UV and visual background radiation. Intergalactic dust can absorb and re-emit this as infrared, contributing to the cosmic infrared background (CIB). Most galaxies seen today have little dust, elliptical galaxies being almost dust free, but ultraluminous infrared galaxies (ULIRGs) colliding or merging with another galaxy involve huge amounts of dust. Such collisions and mergers were more frequent in the past, and the star formation rate of the universe peaked around $Z = 1$ to 2 at one to fifty times the rate seen today. The galaxies in this redshift range contribute 50-70% of the brightness of the CIB.

Chapter 11

Quasars are another source of CIB, matter being absorbed by the central black hole, producing X-rays which are absorbed by the dust torus of the accretion disk and re-emitted in the infrared. This source provides 20 to 30% of the CIB, but is the dominant source of CIB energy at certain specific wavelengths.

Lower energy X-ray background is believed to result from galactic emissions (the galactic X-ray background), higher energy X-ray background coming from a combination of sources outside the Milky Way (the cosmic X-ray background). The galactic X-ray background results mostly from hot gas within 100 parsecs of the sun.

Gamma-ray cosmic background radiation is extragalactic, one of the nearest known pulsars being found to be a source of this form of background radiation.

The cosmic microwave background

The cosmic microwave background (CMB) was discovered in 1964 when Robert Wilson and Arno Penzias were testing a new microwave radio antenna. They could not eliminate all the background noise from their signals and investigated various possible sources of the radiation. Then after speaking to a colleague who knew of earlier predictions of a cosmic background radiation, they realized that it was this that they had observed (Sparrow 2006).

The CMB is sometimes called 'relic radiation' because, according to the Big Bang model of the universe, it was released when the universe had cooled enough to form neutral atoms which could not absorb thermal radiation, thus becoming transparent to electromagnetic radiation, including light. This was about 380,000 years after the Big Bang when the universe was filled with hot, ionized gas. This gas is presumed to have been very uniform, giving rise to the uniformity in the CMB.

When the CMB was initially emitted it would have been mostly visible and UV light, but according to the Big Bang theory it has been redshifted over more than ten billion years to the invisible microwave band. Indeed, this redshifting of 'early light' from distant parts of the universe helps explain Olbers's paradox, which was discussed in the last chapter.

Bumps in the CMB

Slight deviations in the CMB of only about 1 part in 100,000 are, according to the inflationary Big Bang model of the universe, a result of tiny fluctuations of density that arose from quantum mechanical fluctuations after inflationary expansion had eliminated any significant inhomogeneity in the early universe.

These tiny fluctuations then grew with time because of gravitational effects, slightly denser regions attracting more material and increasing in density. Then radiation emitted from denser regions is impeded by higher gravity and undergoes gravitational redshifting (GRS). Thus the apparent temperature map of the CMB is thought to give a map of the density of matter in the early universe.

Temperature of the CMB

The CMB has a thermal black body spectrum with a temperature of 2.725 °K, very close to absolute zero. Its spectral radiance with respect to frequency peaks at 160.2 GHz, the spectral radiance with respect to wavelength peaking at 1.063 mm.

At the 'time of last scattering' (about 380,000 years after the Big Bang) when the CMB was created, the temperature of the universe was 3,000 °K. The temperature of the CMB as a function of redshift, Z, is proportional to its present-day temperature: $T = 2.725(1 + Z)$.

Cosmic dust and CMB

Fred Hoyle argued that fine particles of cosmic dust were prime candidates as possible causes of the lengthening of the wavelength of the background radiation to the microwave band (Hoyle 1983). He notes that much cosmic dust is carbon and that when carbon vapour is cooled slowly, *"long slender needles of carbon are indeed formed, in the process known to laboratory workers as 'whiskering'. The carbon whiskers are just what is needed to give the steady-state theory a chance of being right."*

Hoyle also notes that *"if bacteria really have the universal presence which astronomical observations suggest, I would consider it likely that they are responsible for the microwave background."*

This might be possible because, as noted in Chapter 5, some bacterial species exist in long needle-like forms that conserve water in the arid medium of space. Under favourable conditions, these structures can grow rapidly and are largely made of carbon, the ideal material for generating the CMB.

The Sunyaev-Zel'dovich effect

In the Sunyaev-Zel'dovich effect (SZE) background radiation is scattered and changed in wavelength by collisions with electrons in large clouds of intergalactic gas. Thus the SZE distorts the CMB, observed distortions being used to detect density variations in the universe, and thus observed known clusters of galaxies and detect new ones.

Hartnett (2014) cites a study whose results cast doubt on whether the CMB is from the background. Observing thirty-one relatively close galaxy clusters, most with $Z < 0.2$, an expected decrement in temperature of the CMB as a result of shadowing was only detected in 25% of cases.

The study also looked for an expected temperature change in the X-ray, emitting intergalactic medium via the SZE, but sometimes obtained contradictory results.

Another study of thirty-eight galaxy clusters found that, not only was the SZE less than expected, but it tended to progressively vanish for redshifts from 0.1 to 0.3, casting further doubt on whether the CMB is really from the background (Hartnett 2014).

The SZE can account for the CMB

Proponents of the Big Bang theory hold that the small 'bumps' found in the CMB arise from small variations that resulted during the inflation period soon after the Big Bang, these small variations in density increasing as their greater gravity attracted more matter. Redshifted over billions of years by the expanding universe the CMB has small bumps as a result of these early density variations.

Having failed to do so initially, data from the Cosmic Background Explorer (COBE) satellite was found in 1992 to show these bumps, this being taken by many as final proof of the Big Bang.

An alternative explanation, of course, is the SZE, Sunyaev and Zel'dovich, estimating that bumps as large as one part in a thousand could occur as a result of the SZE. Another author estimated three parts in ten thousand, whilst the COBE observed only about one part per hundred thousand.

Thus the 'original' CMB might be perfectly uniform in temperature and the SZE could easily account for the 'bumps' found in it when it reaches Earth. If the CMB were from a relatively smooth source that would contradict the Big Bang theory (Humphreys 2014).

Hydrogen interactions and the CMB

David Noel argues that the CMB is "merely the interchange of microwave quanta between the atoms of interstellar and intergalactic hydrogen."

He notes that enclosed gas molecules transfer energy by collisions, the average distance between collisions being termed the mean free path. For interstellar gas, the mean free path is extremely large and collisions are infrequent, but equilibrium can be maintained by the interchange of microwave quanta (Noel 2014).

He also argues, "An expanding universe would not, of itself, stretch wavelengths of radiation travelling within it. If it did, then all radiation from distant galaxies billions of light years away, and hence billions of years old, would also be stretched down towards microwave length, very much more than in observed redshifts."

Conclusions

That there is a spectrum of background radiation is to be expected. As noted in the first section of this chapter, the sources of infrared, X-ray, and gamma-ray background radiation are now known, for example 70% of cosmic infrared background radiation being emitted by galactic dust particles that re-emit absorbed UV and visible light as IR.

Similarly, interstellar dust is perhaps the most likely source of the CMB (Hoyle 1983).

Hoyle also proposes interstellar bacteria as another possible source of the CMB, but the present authors think cosmic dust, of which there is a good deal, as noted in Chapter 5, is the more likely source of the CMB.

On a universe-wide scale, the distribution of cosmic dust would be relatively uniform, accounting for the uniformity of the CMB.

Another possible source of the CMB is the SZE, and this would easily account for the small bumps observed in the CMB.

Noel's attribution of the CMB to interchange of microwave quanta between hydrogen atoms in the interstellar medium is, perhaps, another possible mechanism by which the CMB could be created.

As a bottom line:

(a) having so many varied sources of radiation, some of them immensely more intense than our sun but extremely distant, the universe should be expected to have some form of background radiation, and any 'background' radiation would be expected, of course, to be relatively uniform.
(b) interactions with interstellar cosmic dust seem a sound enough explanation for the CMB, whilst other mechanisms such as the SZE, and interchange of microwave quanta between hydrogen atoms in the ISM, might also contribute to it.
(c) the slight 'bumps' found in the CMB are clear evidence that it did not originate in some homogeneous fashion from a hypothetical Big Bang, but that a few larger but distant galaxies have higher emissions of the radiation that is shifted to the microwave region of the spectrum.
(d) with dark matter and dark energy as yet unproven, and the Higgs boson detected with some uncertainty only recently, it could also be proposed that the CMB may be, at least in part, caused by such universal entities as some form of dark matter such as 'cold dark matter'. It does seem more likely, however, that the CMB arises from a present day source that is spread throughout the universe with great uniformity, whatever that source may be.

Chapter 12

Scientific Theories of the Universe

> *The big bang today relies on a growing number of hypothetical entities, things that we have never observed—inflation, dark matter, and dark energy are the most prominent examples. Without them, there would be a fatal contradiction between the observations made by astronomers and the predictions of big bang theory.*
>
> E Lerner, Bucking the big bang, *New Scientist*.

Introduction

The following chapter discusses 'scientific theories' of the universe, that is, various theories proposed by the scientific community though some of these may seem rather 'hair brained' to most readers.

First, therefore, the steady-state and Big Bang theories are summarized, having been discussed at some length in Chapters 7 and 8.

A number of alternative scientific theories of the universe are then discussed, whilst religious theories are then discussed in Chapter 13.

Finally, the evolving universe theory favoured by the authors is discussed in Chapter 15.

Chapter 12

The steady-state theory

The steady-state theory has been discussed at modest length in Chapter 7. This is, of course, the most 'natural' theory. In other words, given that we know that there has been life on our planet for millions of years, and that the universe is billions of years old, it seems reasonable to take the Earth we live in, and the universe around it, as a 'given' that is here to stay.

Clearly, however, the universe is not in a steady state, and massive changes occur at various places in it from time to time, including formation of new stars and collapse of old ones. Steady-state theory took note of such changes in proposing creation-fields that produce new matter to keep the density of the universe relatively constant as it expands.

With the discovery of the cosmic microwave background radiation (CMB) and its attribution to the 'time of last scattering' soon after the big bang, the Big Bang theory overtook the steady-state theory in popularity.

There have been claims, however, that 'static' models of the universe predict the CMB better, for example:

We present the history of estimates of the temperature of intergalactic space. We begin with the works of Guillaume and Eddington on the temperature of interstellar space due to starlight belonging to our Milky Way galaxy. Then we discuss works relating to cosmic radiation, concentrating on Regener and Nernst. We also discuss Finlay-Freundlich's and Max Born's important research on this topic. Finally we present the work of Gamow and collaborators. We show that the models based on a Universe in dynamical equilibrium without expansion predicted the 2.7 °K temperature prior to and better than models based on the Big Bang (Assis and Reeves 1995).

The Big Bang theory

The Big Bang theory holds sway at present because it is believed to better explain

   [1] redshift of light from distant stars and galaxies.
   [2] the very uniform CMB and its black body properties.
   [3] the relative abundance of lighter elements in the universe.

Scientific Theories of the Universe

The Big Bang theory, however, has had to be adjusted several times, including

- inclusion of an inflationary period to account for the flatness of the universe and the uniformity of the CMB.
- luminosity correction for galactic evolution to explain failure of the Tolman test.
- dark matter to explain galactic rotation curves and the dynamics of clusters of groups of galaxies.
- dark energy to explain the acceleration of distant supernovas which is taken as evidence of accelerating expansion of the universe.

As a result, thirty-three leading scientists have published an open letter to the scientific community on the Internet at www.cosmologystatement.org, and in the journal *New Scientist* (Lerner 2004). An extract from this letter is quoted at the beginning of the chapter, and the letter goes on to say such things as:

The big bang can't survive without these fudge factors.
What is more, the big bang theory can boast of no quantitative predictions that have subsequently been validated by observation.

One blow for the Big Bang theory was the finding that the small anisotropies in the CMB attributed to inflation after the Big Bang could be attributed to the SZE, that is, to scattering of radiation by electrons in the interstellar medium. Another was that expected shadowing of the CMB by thirty-one relatively nearby galaxies was, on average, not found within the margins of experimental error (Hartnett 2006).

Cyclic universe theories

Whilst originally his own model of the universe was static, Einstein's general theory of relativity encouraged ideas of dynamical models of the universe and in the 1920s Alexander Friedmann and Georges Lemaitre showed that many models of the universe could be based on the assumption of space curvature changing with time.

Friedmann's 1923 cyclical model of the universe had a period of about 10^{10} years, based on the assumption of a zero cosmological constant and the mass of the universe being 5×10^{21} solar masses.

In 1933, Lemaitre considered cyclic models of the universe but concluded that they were not supported by observation.

The quasi-steady state cosmology (QSSC) was an extension of the steady-state theory that attempted to explain the surplus of faint radio sources and the evolution of the CMB by proposing that the growth rate of the universe varied sinusoidally over time but with a diminishing amplitude of variation (Hoyle et al. 1993).

The period of these oscillations was taken to be about three times the age of the Big Bang universe, each cycle beginning with a *little bang* and the universe currently being in an expansion phase being the explanation for redshift.

It was claimed that the QSSC model could explain the relative abundance of lighter elements in the universe just as well as the Big Bang model:

> *A discussion of the origin of 2D, 3He, 4He, 7Li, as due to the x-process in primary nucleosynthesis. While it is generally argued, following Gamow, Alpher, and Herman, that these isotopes were synthesized in a Big Bang, it is equally likely that these isotopes were made in active galactic nuclei, as was the cosmic microwave background (CMB), in a cyclic universe model* (Burbridge at al. 1957).

Another cyclic theory is that of the pulsating universe, which is assumed to expand for a certain period of time and until it reaches some upper size limit. It then undergoes a period of contraction and when a certain lower limit of size is reached expansion resumes again.

Other cyclic theories propose a Big Bang and expansion of the universe until it collapses with a 'Big Crunch', this being followed by another Big Bang, and so on.

In the conformal cyclic cosmology (CCC) model the universe iterates through 'infinite' cycles, with the future "timelike infinity" of each previous iteration being identified with the Big Bang singularity of the next (*Wikipedia* 2014).

Each sector of this model is called an 'aeon' and, in line with the second law of thermodynamics, involves gradual loss of entropy and associated increases in 'randomness' (Penrose 2010).

Claims that concentric anomalies found in the CMB were consistent with the CCC model have since refuted by findings that these anomalies were statistically insignificant.

Multiverse theories

One of these proposes that the inflation process of the Big Bang occurred repeatedly to generate a multitude of universes or a 'multiverse'. It then proposes that random fluctuations in the dark energy field are the cause of new rounds of inflation which create further universes.

Some of the many universes are thought to vanish by collapsing into black holes and evaporating, while others expand until they thin out to become a new almost empty space in which further inflation can occur.

One multiverse theory proposes that time is reversible to the extent that some of the many universes move backwards in time.

The multiverse theory proposed by physicist Andrei Linde is that our universe is one of many that grew from a multiverse consisting of vacuum that had not reached its ground state. In this scenario the universe "tunnelled" from pure vacuum to form a bubble of 'false vacuum' with no matter or radiation. The space inside this bubble was curved and it contained a small amount of pent-up energy, this violation of energy conservation being allowed by the Heisenberg uncertainty principle for small time intervals. The bubble then inflated exponentially in a manner comparable to inflation after the Big Bang, its energy being converted into matter and radiation. Then as the universe cooled, it took on the structure we see today.

Another multiverse theory is the ekpyrotic scenario which originated from the work of Neil Turok and Paul Steinhardt, the term ekpyrotic being Greek for 'conflagration.' This model used string theory to postulate that the Universe was but one of a number of 3-D 'branes' (short for membrane) moving through a 4-D background called 'the bulk'.

When two of these branes, having expanding greatly, collide, this can result in an explosion and subsequent inflation similar to the Big Bang, but results in slightly different variations in the CMB and a different frequency distribution of the gravitational wave spectrum.

The ekpyrotic model is cyclic but collisions between branes are rare, being supposed to be of the order of a trillion years apart as the universe expands into an almost featureless flat expanse (Frank 2008).

A variation of this model proposes that the bulk universe has 4-D stars and that a supernova-type explosion of a massive 4-D star created a black hole. The boundary between the inside and the outside of this 4-D black hole, or its 'event horizon', would be a 3-D 'hypersphere' containing our universe.

Plasma cosmology

In 1965, Hannes Alfven proposed a 'plasma cosmology' theory of the universe based on lab experiments and scaling observations of space plasma (ionized gas with free roaming electrons that can conduct electricity).

Alfven and Oskar Klein suggested that, since most of the local universe was composed of matter, there may be large 'bubbles' of matter and antimatter elsewhere that would, overall, balance to equality. Such bubbles would sometimes overlap, however, when matter-antimatter annihilation would quickly produce high energy photons, but we would not see these because their origin would be outside the observable universe.

Like steady-state theory, plasma cosmology assumes the large-scale universe to be isotropic in time as well as in space, matter having always existed, or at least being formed so far in the past that its origins are beyond investigation.

The increasing mass theory

Another theory in opposition to that of expansion of the universe is Christian Wetterich's proposal that the atoms in the universe are becoming heavier, so that light from lighter, less energetic atoms of the past, would be redshifted (Wetterich 2013).

The basis for this theory is that the characteristic light emitted by atoms is governed by the masses of their subatomic particles, particularly the electrons. Thus light from less heavy atoms of the past would be less energetic and thus of lower frequency and greater wavelength.

According to Wetterich's model, the current cosmos could be static, or even contracting. The universe still expanded very rapidly during a short period of inflation, but prior to inflation the Big Bang stretches out in the past over an effectively infinite period of time.

The great problem with this model is that it can't be tested because mass is a defined 'dimensional quantity' which can only be measured on a relative scale. Thus, if the mass of everything were changing, then the mass of the standard kilogram kept by the International Bureau of Weights and Measures in Paris would also change.

Wetterich's model avoids the notion of a singularity just before the Big Bang when all the known laws of physics break down, and instead, the universe is always in a state of flux with no beginning and no end. Just as cyclic models often propose huge time frames that cannot be tested, this model is purely 'guesswork', highly improbable, and of no real use, contributing nothing new to our understanding of the cosmos.

The Big Freeze model

A group of theoretical physicists in Melbourne have proposed a 'Big Freeze' model of the universe. This compares the universe to a liquid which, as it cools, "crystallizes" into the three spatial dimensions and one time dimension that we know today.

This model is based on the theory of 'quantum gravity' proposed by Canadian physicists in 2006, in which space-time is like a lattice constructed of discrete space-time building blocks.

Chapter 12

According to the Big Freeze analogy, there should be 'cracks' between the different 'blocks' of the universe at some level, whether this be microscopic or on a scale of light years, and researchers hope to find evidence of these cracks.

The placid universe theory

As noted in Chapter, 9, David Noel points out that measured redshifts from distant galactic objects show dispersion, that is, they vary with wavelength, whereas the Doppler effect does not involve dispersion. He therefore claims that gravitational effects are largely responsible for the redshifts attributed to recession velocities and the Doppler effect (Noel 2012).

In another article, Noel attributes the CMB to interchange of microwave quanta between hydrogen atoms in the interstellar medium (Noel 2014).

Noel concludes that there was no Big Bang and that the universe is relatively static, though undergoing substantial change on a smaller scale, an example this being the substantial changes in climate that the Earth is believed to have undergone over recent millennia. Thus he proposes "the placid universe model" which is a return to the original steady-state models without creation fields to maintain constant density of matter in an expanding universe (Noel 2014b).

Conclusions

The Big Bang model of the universe has been amended several times to explain new findings such as quasar velocities near that of light. As a result there is growing dissatisfaction with this model, some feeling that quasars may in fact be far closer and slower than their redshifts suggest, casting doubt on Hubble's aw and thence the Big Bang theory.

Cyclic models of the universe, including the quasi-steady state cosmology, have little to offer, their periods of oscillation usually being assumed far to great to ever be observed in any way.

Multiverse models take matters from the highly questionable to the realm of the ridiculous, some still requiring dark energy, which is unlikely to ever be found, others using string theory to arrive

Scientific Theories of the Universe

at absurd proposals such as 'branes', spaces in which alternative universes exist.

Models involving branes propose that our universe resulted from a collision between two branes, or that a 4-D brane collapsed to form a black hole whose event horizon gave rise to our universe.

Plasma cosmology models propose 'bubbles' of matter and antimatter. These too are, at best, guesswork stuff without any real substance.

Wetterich's 'increasing mass' model is more of the same and again, it cannot be verified because it has a 'relativistic' basis.

The Big Freeze model models the universe has having cooled to such an extent that is can be compared to a 3-D sea of ice, which is supposed to be comprised of discrete 'building blocks.' This theory is yet another 'university staff party' exercise which has little substance.

Noel's 'placid universe' proposal of a steady-state model without expansion brings us back to square one.

Clearly, however, the universe is seething with activity, massive stars involve unimaginable amounts of activity, new stars and galaxies are still being created occasionally, and countless meteors and other space debris constant fly hither and thither. In Chapter 15, therefore, we propose the evolving universe model, one more in keeping with what we know of the universe at present, whilst religious beliefs about creation are discussed in the following chapter.

Chapter 13
Religious Beliefs

> *In the beginning of creation, when God made heaven and earth, the earth was without form and void, with darkness over the face of the abyss, and a mighty wind that swept over the surface of the waters. God said, 'Let there be light', and there was light; and God saw that the light was good, and he separated light from darkness. He called the light day, and the darkness night. So evening came, and morning came, the first day.*
>
> Genesis 1.1, *The New English Bible*.

The development of religions

Until the agricultural revolution of only about 12,000 years ago, hunter-gatherers lived in small groups in which the elders passed on tribal mythology and beliefs. A central belief was that the thinking being within a man was an eternal 'spirit', leading to *ancestor veneration*. As language developed and words were ascribed to a confusing array of objects and phenomena, people began to associate spirits with these objects and *animism* and *nature worship* developed.

In those times, people lived in circumscribed areas, and there was only a limited number of tales of folklore and superstition to be told and passed on. Furthermore, without radio and TV, et cetera, there was little else to do during a long winter evening beside a communal fire but to listen to these tales or talk about the events of the day.

Thus tribal beliefs were passed on from generation to generation and, indeed, a few isolated tribes in remote areas such as the New Guinea highlands still have traditional religions that have changed little over time.

Generally, however, as man came to live in larger communities of perhaps a hundred or more people, more formalized religious beliefs developed and shamans were chosen to expound them and conduct rituals in specially appointed places, these sometimes including a special hut built for the purpose. Without shops, pubs, or clubs to go to, of course, there were no other venues to compete with religious sites for 'entertainment', which, of course, religious rituals, sermons and so forth were, at least in part.

This specialization of occupations increased with the advent of the agricultural revolution, after which small towns were built, and eventually small cities, when the first small temples were built, and priests appointed to run them.

Most primitive polytheistic religions had some sort of hierarchy of deities, and the traditional religions of New Guinea believe in a supreme creator God.

It was not until about 2,500 years ago, however, that the roots of monotheism were established in Hebrew folklore, in particular that of Jacob: son of Isaac, brother of Esau, and father of the twelve patriarchs of Israel. Legend has it that Jacob wrestled with God who then named him Israel, meaning 'one who has been strong against God'. As with more ancient polytheism, we see here folklore being the basis of religion.

Here, the God of the Hebrews (Yahweh, or Jehovah) seemed to serve a political purpose, principally that of making the Jews the 'chosen people' who he would free from their bondage in Egypt and deliver to the 'Promised Land'. Notably, Moses, who had been brought up by a daughter of the Pharaoh, was 80 and living in exile when Yahweh commanded him to lead his people to the Promised Land (Mohr and Fear 2014).

Here, then, we see a story that repeats itself often in history, namely of a person of sometimes noble upbringing being in the wilderness and receiving a message from God.

Thus the person is able to establish a religion, as Moses supposedly did with the Ten Commandments after another 'session with God' on his own, without witnesses of course.

Other comparable stories are those of Zoroaster, Buddha, Guru Nanek (Sikkhism), Joseph Smith (Mormonism), and Wallace Fard (Nation of Islam). They too all claimed visions of, and messages from God, to which there were no witnesses (Mohr and Fear 2014).

Comparable stories too, insofar as they were based on visions and conversations with God, were those of Jesus and Muhammad, both of whom claimed God had made them prophets. In both cases these prophets seemed bent on replacing an existing religion with a new one, probably for political purposes.

In similar fashion, several of the many Protestant sects of Christianity were born out of political concerns about the power of Rome.

Thus it was that countless religions have been created by fairy tales and obvious lies, often for political purposes, and therefore often creating much conflict. Indeed, it seems that the extreme fundamentalists of the present time should be a reminder that those who preach religion are often, if not usually, evil people following a hypothetical but vengeful and evil God. As always, however, such activity is an occupation which provides income, power and influence, as evidenced by the often enormous temples and churches built as an exercise of that power (Mohr and Fear 2014).

Why are we so gullible as to let them get away with it? The same question is easily asked in today's 'consumer society' in which we are brainwashed *consumer zombies* with a bottle of soft-drink or beer in one hand, a cigarette or mobile phone in the other, wearing often ridiculous clothes, and driving a 4-wheel drive car to the supermarket because it has become 'fashionable.'

Just how we become consumer zombies is discussed at length in the first author's book *The Pretentious Persuaders* (Mohr 2013b).

Creationism

Creationism is the term used to summarize religious beliefs that a supernatural being or god created the universe, an example of such belief being the quotation from the New English Bible that opens this chapter.

As the development of science became much more rapid during the eighteenth century, bringing with it Europe's Industrial Revolution, attempts were made to reconcile the Abrahamic and Genesis creation stories with science.

In the nineteenth century, Darwin's theory of evolution by natural selection divided opinion greatly, and the term 'anti-evolutionists' became common.

In the United States in 1920s, the term 'creationism' was first associated with Christian fundamentalists, and later with others such as 'evolutionary creationists'. Since then literalist creationism in the United States has contested scientific theories such as those of evolution and the Big Bang, taking particular exception to the notion of 'common descent'.

According to *Wikipedia*:

Today, the American Scientific Affiliation, a prominent religious organization in the United States, recognizes that there are different opinions among creationists on the method of creation, while acknowledging unity on the Abrahamic belief that God "created the universe".

Intelligent Design

Intelligent Design (ID) is a modern version of creationism which holds that some features of our world and the many life forms in it are too 'ordered' and complex to have been a result of such relatively random processes as natural selection, and must therefore have resulted from 'design' by a higher being.

The ID movement was developed by a group of American creationists to circumvent US court rulings against compulsory teaching of creationism in public schools on the basis of breaching separation of church and state.

According to *Wikipedia*, the leading proponents of ID are supported by the Discovery Institute, a politically conservative think tank based in the United States, and they believe the 'designer' to be the Christian deity.

The first publication of the phrase *intelligent design* with its present use as an alternative term for creationism was in a 1989 high school biology textbook. In 2005, a US district judge ruled that ID is not science and that it "cannot uncouple itself from its creationist and thus religious antecedents", and that promotion of ID therefore violated part of the First Amendment to the US Constitution.

Fine-tuning and alien design

Some have proposed that the universe might have been designed by extra-universal aliens. The basis for such beliefs is that the universe, at least in part, is 'fine-tuned' to provide the conditions necessary to allow life to exist. Indeed, cosmologist Alan Guth is reported to believe that one day humans will be able to generate new universes.

Such ideas imply that extraterrestrial designers were themselves a product of evolutionary processes. Thus the designer universe theory of John Gribbin suggests that the universe could have been made by a technologically advanced civilization in another part of the multiverse.

Some theologians argue that the exceptional conditions necessary for life are unlikely to have resulted by chance and that God must have been responsible for fine-tuning.

Philosopher and Christian apologist William Lane Craig cites fine-tuning of the universe as evidence of God or intelligence capable of manipulating the fundamental laws of physics. He concludes, however, "that the postulate of a divine Designer does not settle for us the religious question" (*Wikipedia* 2014).

Conclusions

Mankind's many religions have evolved considerably, from animism to polytheism, and finally to monotheism. There having been a considerable reduction in the number of spirits and gods that we believe in, therefore, it is not surprising that agnosticism and atheism are increasingly common whilst many religions are waning in popularity in a modern world that provides many other diversions.

Intelligent Design was an attempt by creationists to counter the increasing religious scepticism that advances in science such as Darwinism had caused.

Religious Beliefs

Perhaps the bottom line on religion, however, is that *we* invented gods and the many fictitious stories that surround them, not the other way around. Thus what we really need is self-belief and self-improvement, not religions in the name of which countless millions have been slaughtered for millennia.

As for alien design, that is the stuff of sci-fi movies, and should not be taken seriously. Likewise suggestions that the universe is a computer simulation by future renditions of ourselves or aliens.

The penultimate chapter of this book discusses the evolving universe theory in which the universe is seen to be evolving into a relatively steady state.

Chapter 14

Is There Life Out There?

> The structure of Pedomicrobium is so distinctive that there can be no possibility of mistaking it, and when Pflug found carbonized examples of this bacterium—indeed whole clusters of it—in the meteorite, the issue which had been so controversial swung in favour of the claim of Claus and Nagy, which had been shouted down so vociferously twenty years before. Here surely is clear evidence of extraterrestrial life.
>
> Fred Hoyle, *The Intelligent Universe*.

Traces of bacteria in meteorites

In 1864, a shower of meteorites fell near Orgeuil in southwest France. Examination of sections of these meteorites in the early 1930s found them to contain a significant amount of carbon as spherical skins surrounding grains of inorganic material, raising the possibility that this material was coalified bacteria.

Coalification occurs in terrestrial rocks when the outer cell wall of a bacterium or spore is transformed into carbonaceous material, the less durable matter inside eventually being reduced from organic to inorganic form. The Orgeuil specimens appeared likely to have undergone this process because their skins were smooth and often double like biological cells (Hoyle 1983).

In the early 1960s, George Claus and Bart Nagy found other intriguing structures in Orgeuil specimens and also in another meteorite that had fallen in Tanzania in 1938. They found filamentatious skins resembling microscopic fungi and objects they called "organized elements", which they concluded had been living organisms.

Radioactive dating of these meteorites found them to be as old as the solar system so that they predated Earth itself slightly and was proof that life in space predated that on Earth. For the most part, however, the scientific community refused to accept that meteorite specimens were proof of life in outer space.

Then in 1969 a carbon-bearing meteorite was found near the town of Murchison in Victoria, Australia. Ten years later Hans Dieter Pflug began studying it, finding structures similar to those found in France and Tanzania.

For his studies, Pflug dissolved away the mineralized portion of a thin slice of meteorite with acid, leaving a carbon-bearing residue, which was examined with an electron microscope. Pflug found tiny filaments very similar to fossils found in Gunflint chert in northern Minnesota, which had been widely accepted by palaeontologists as being of biological origin.

Then in 1981, Pflug found clusters of carbonized Pedomicrobium, a bacterium with an unmistakable flower-like appearance which 'feeds' on metal compounds by oxidizing them. According to Hoyle (1983),

Here surely is clear evidence of extraterrestrial life.

Some of Pflug's collection of electron micrographs also showed the characteristic hexagonal patterns of the outer cell walls of the class of bacteria called methanogens.

The origin of life on Earth

Hoyle argues that Pedomicrobium may have been responsible for oxidization of ferrous iron in the sea up to 3,800 million years ago before there was sufficient oxygen in the atmosphere to do so, which was not until about 2,000 million years ago.

Chapter 14

If this early Pedomicrobium came from space, this leaves an unexplained gap of 700 million years back to when the Earth was formed about 4,500 million years ago.

During that 700-million-year period, the solar system orbits still overlapped somewhat and violent collisions between major objects in it occurred. These were not conditions suitable for evolution of the first bacterial life on Earth to emerge from some primordial soup, but an extreme form of the sort of situation in which the dinosaurs and all other larger life forms became extinct in the Mesozoic era between only 63 and 230 million years ago.

Thus Hoyle argues (1983):

> *Furthermore, the geological history of the Earth supports the idea that single-celled creatures like Pedomicrobium did not evolve, but arrived suddenly as soon as conditions were tolerable.*

The most likely source of life from space is comets, of which there are billions, and which are in a hard-frozen state ideal for preservation of organic material.

Examination of the trace material left by evaporation of parts of comets as they move close to the sun has shown them to contain far more carbon and nitrogen in proportion to hydrogen and oxygen than does the biosphere. Indeed, the proportions of these elements in comets are closely similar to those found in bacteria and mammals.

Thus comets may provide a possible means of transmission of life within and between the hundreds of millions of galaxies in the universe, sheer numbers allowing the possibility of evolution of advanced life forms as on Earth.

For most of their life-span of millions of years, comets have orbits far from the sun, but when some pass close to planets, they are pulled into a smaller orbit closer to the sun. Some of these short-period comets then collide with asteroids in the asteroid belt, some of the resulting fragments then falling on planets such as Earth as meteors.

Before they fall on Earth, meteorites orbit the sun millions of times, alternately being roasted and frozen so that any initially alive material in them is coalified, leaving the fossil evidence found in the Murchison and Orgeuil meteorites.

Bacteria in the interstellar medium

Fred Hoyle and Chandra Wickramasinghe began studying the composition of interstellar dust in the 1960s.

In 1968, polycyclic aromatic molecules were detected in interstellar dust, and in 1972, convincing evidence that the dust contained porphyrins was obtained.

In 1974, Wickramasinghe proved that molecules of poly-formaldehyde, which is closely related to cellulose, were in the interstellar medium.

As noted in Chapter 5, in 1979, Hoyle and Wickramasinghe compared the interstellar extinction curve with that for dried bacteria, obtaining a very close match (Hoyle and Wickramasinghe 1985). They concluded that the extinction curve was probably caused by dried, frozen bacteria.

This result has not been widely accepted, but Hoyle (1983) also pointed out that such bacteria could be responsible for the cosmic microwave background (CMB) radiation usually attributed now to the Big Bang.

Indeed, some bacterial species exist inside small needle-like sheaths which conserve water in the harsh medium of space, the ideal needle shape required to transform starlight into radio waves.

That bacteria can survive in space was spectacularly demonstrated when in 1969, the Apollo 12 astronauts recovered a TV camera that had been left on the moon by the unmanned lunar lander *Surveyor 3* two and a half years earlier. When it was examined back on Earth, *Streptococcus mitis* were found still alive on it. NASA determined that, because of the precautions the astronauts who retrieved the camera had taken, the germs inside the camera must have come from Earth via the unmanned lunar lander (Klyce 2014).

Survival in such harsh conditions with massive temperature swings and no water is possible because bacteria can become 'micro-cannibals', most bacteria dying to supply the rest with

Chapter 14

protective proteins, gums, sugars and cryoprotective substances. For example, bacteria seem to have survived 4,800 years in the brickwork of Peruvian pyramids, 11,000 years in the gut of a well-preserved mastodon, and as much as 300 million years in coal (Klyce 2014).

The ultimate survival mechanism for bacteria, however, is formation of spores with protective coats that become completely dormant. Some bacteria form endospores, the original cell replicating its chromosome and surrounding one copy with a durable coating which can survive in boiling water for long periods of time.

In 1995, two biologists extracted bacterial spores from bees preserved in amber in Costa Rica that was between 25 and 40 million years old. When placed in a suitable culture the spores returned to life. On learning of this discovery, John Postgate, author of the book *The Outer Reaches of Life*, wrote:

> 'Could life on this planet be descended from alien spores? . . . Panspermia, the view that the seed of life is diffused throughout the universe, has been favored by a minority of thinkers since the Greek Anaxagoras in the 5th century BC. He, Arrhenius and Fred Hoyle may yet have the last laugh on us doubters.

Can bacteria survive the journey to Earth?

Factors which increase the chances of bacteria surviving the temperatures generated by entering the Earth's atmosphere include:

[1] bacteria are far smaller than 'shooting stars' and would not reach such high temperatures and burn up.
[2] particles entering the atmosphere tangentially as returning astronauts do would be heated much less than those entering 'head on'.

This still leaves the problem of bacteria having to survive high temperatures for a few seconds but bacteria in hot volcanic chimneys on the sea-floor near the Galapagos Islands have been found to cope with a temperature of over 300 °C. This result in hot liquid water suggests that dry bacteria could survive at least 200 °C for a short period.

On this basis, Hoyle calculated that microorganisms up to about 0.1 mm in diameter could survive passing through the upper atmosphere at 32,000 km/hr, and this size is large enough to host colonies of bacteria (Hoyle 1983).

Can bacteria survive in space?

The survival of bacteria left on the moon for two years by an unmanned lunar lander was noted earlier. The main threat to microorganisms in space is low-energy X-rays from the stars and objects outside the Milky Way. Such radiation can permanently damage the DNA of cells, but as noted above bacteria are often able to repair such damage.

In one experiment, the bacterium *Micrococcus radiophilus* was subjected to a large dose of X-rays that caused 10,000 breakages in its DNA, but the bacterium was able to repair this damage and become viable again (Hoyle 1983).

A species of the *Pseudomonas bacterium* was found living inside an American nuclear reactor in 1960, having suffered radiation exposure millions of times greater than has existed on Earth during any period during which life could have survived here, that is for the last 3,800 million years.

Evolution on Earth

Increasingly, many scientists see the likelihood of life evolving on Earth 'from scratch' as improbable. This is because classical Darwinian evolution requires a long sequence of DNA copying errors to produce a new species that is a variation of a pre-existing species.

Some understanding of this process is gained by reviewing the basic stages of cell reproduction which are

S: DNA copying which takes 6-8 hours.
G2: preparation for cell division which takes 3-4 hours.
M: cell division (mitosis) which takes about an hour.
G0: sleeping or quiescent mode for days up to years.
G1: preparation for the next DNA copy is initiated when cyclin-dependent kinases (CDKs) are guided by cyclins to attach phosphate ions to proteins which cause cell copying, and to other proteins which cause cell division.

Mutation is often the result of a carcinogen-induced, virus-induced, or random DNA copying error. Random DNA errors are partly responsible for evolution and are just a statistical fluke, having a probability of one in a million or less, but this amongst billions of cells and many cell cycles (Mohr 2013a).

In contrast, to produce life 'from scratch' successfully, however, requires an almost impossibly long sequence of error-free DNA copying. Suppose, for example, a simple protein requires ten links in the DNA in the correct sequence.

To obtain the correct ten links by the known rate of miscopying would required about a hundred thousand billion copying operations. Thus, given 100 million members of a species all reproducing, it would take a million generations to produce the required sequence.

Such improbable numbers lend support to the cosmic theory of the origins of life on Earth.

Other microorganisms

Bacteria are believed to be the microorganisms from which life evolved, whether these originated on Earth or came from space. Several other microorganisms are involved in this evolution, however, including:

[1] diatoms, microscopic water plants which form the bottom of aquatic food chain and generate much of the planet's oxygen.

Is There Life Out There?

[2] micro fungi such as yeast.
[3] protozoa such as the amoeba.
[4] viruses, the largest being of comparable size to the smallest bacteria. These reproduce by inhabiting larger biological structures, tiny genetic fragments called viroids containing only one or two genes.

Unlike these microorganisms, large multicellular animals cannot survive at the very low pressures and high temperatures of outer space, suggesting their different origins. It is possible, however, that viruses from space might have contributed to evolution. This can happen because viruses and viroids do not always multiply when they take over host cells, instead they sometimes add their own genes to host cells.

That much of the DNA of all plants and animals is redundant is perhaps evidence of such processes, 95% of human DNA being redundant, even higher percentages of redundant DNA occurring in some lower plant and animal species.

DNA that is redundant in one species, however, may be active in another, those which produce blood in animals being inactive pseudogenes in plants, emphasizing the way in which all life forms have developed as parts of an 'evolutionary tree' of life. Further evidence of this common background is the way in which chemicals from plants and fungi such as morphine, quinine, and penicillin interact strongly with humans. A more notable example is that the juice of the king-coconut is interchangeable with human blood plasma.

Hoyle argues that acquisition of genes from external sources (i.e., space) fits the known 'occasional jumps' nature of evolution, an appropriately ordered group of new genes being required before each evolutionary step (Hoyle 1983).

Viruses from space would have to arrive within bacterial hosts, of course, and be able to cope with the survival mechanisms of bacteria. A study in Finland induced bacteria which encapsulated viruses to form spores. When the spores were revived into active bacterial colonies some of the spores still contained viruses, showing that viruses could indeed 'hitch a ride' through space to Earth (Thomas 2014).

Chapter 14

Epidemics from space

Not only may genetic material from space have contributed to evolution, but new genetic material may still be reaching Earth all the time, and there has occasionally been "clear evidence" of epidemics of such diseases as influenza having arrived from outer space.

A possible example of this was found in a study of the incidence of the common cold in different parts of the Netherlands in 1925-26 which found that hill-shaped infection rate graphs were much the same in different parts of Holland.

Another is that infection rates for Hepatitis A over an eight-year period in New York State were twice as great in urban areas as in the crowded city, and four times as great in sparsely populated rural areas.

Similarly, two studies of influenza infection rates found that there was no significant increase in infection rates within households, compared to within the community at large, suggesting that the virus "fell vertically" (Hoyle 1983).

Most scientists doubt the possibility of viruses falling to Earth from space and infecting us or mixing with our DNA. It is possible, however, that viruses could have contributed to the evolution of life on Earth by finding their way here via host bacteria within meteorites or in interstellar dust.

Is there life on Mars?

In 1976-77, NASA's *Viking I* landing carried a Labelled Release (LR) experiment and an organic analysis experiment (GCMS) designed to detect signs of living matter. The LR experiment proved positive, but the other negative, though Hoyle argues that both experiments should have had a control test under Antarctic soil conditions.

In any event, subsequent attempts to reproduce the positive LR result using sterile soil samples have always failed, reinforcing the significance of the first positive result on Mars.

Indeed, Hoyle argues that Mars' redness is a sign of abundant oxygen in its past that could have only come from such processes as photosynthesis or from iron-oxidizing bacteria such as

Pedomicrobium. Like many other scientists, he also feels that life might still be present in liquefied parts of the many glaciers on Mars, and that the surface of Mars "looks very much like a failed attempt at seeding life from space, a failed 'experiment' of a kind which eventually succeeded in the case of Earth" (Hoyle 1983).

In 1984, a two-kilogram meteorite was found in Antarctica's Allen Hills. It has broken away from the surface of Mars 16 millions years ago, reaching Earth about 13,000 years ago. Electron microscope studies showed wormlike structures similar to those of fossilized bacteria found on Earth, along with traces of magnetite, a chemical normally manufactured by bacteria on Earth (Sparrow 2006).

Bacteria and space research

An example of bacterial survival in space was provided by those that survived for two years on a camera left on the moon. An example of how that situation occurred is provided by findings that the previously undiscovered bacterium Tersicoccus phoenicis can survive in NASA and European Space Agency (ESA) clean rooms. These clean rooms have only a few hundred microbes, far less than anywhere else on Earth, but this new species was found in clean rooms 2,500 miles apart (Shaunacy 2013).

In 2010, and again in 2011, NASA studied the bacterium *Pseudomonas aeruginosainto* cultured in artificial urine aboard the Atlantis shuttle, finding that in space the bacterial colonies or 'biofilms' were thicker than those grown on Earth, also having a previously unseen "column-and-canopy" structure, suggesting that lesser gravity favours bacterial growth (Kramer 2014).

Conclusions

The first bacteria that colonized earth from space some 3.8 million years ago encountered hostile conditions. There was no free oxygen and no ozone to block UV radiation from the sun. Instead the air was hot and filled with sulphurous volcanic gases and nuclear radiation from U235 was 50 times greater than now.

Chapter 14

Anaerobic, alkophilic, and sulphur bacteria, however, can survive such conditions, as bacteria found today in deep ocean trenches and hot springs demonstrate.

In addition, bacteria have highly efficient survival mechanisms, for example being able to regenerate after being frozen (Klyce 2014)

If the first life on Earth was indeed bacteria from space, where did that originate? One possibility, of course, is Mars, which it is thought was once more hospitable to life than now, and which may still have surviving microorganisms in liquid parts of its many glaciers.

Indeed, NASA having landed unmanned craft on Mars, its Planetary Protection Officer Catherine Conley is thinking ahead so that future manned missions will have the best possible precautions to prevent dangerous microorganisms being brought back to Earth (Biba 2012).

As for evolution, life forms required a stable liquid such as water in which organic chemicals can dissolve and move around, contacting other molecules and forming more complex structures as they do so. In the solar system liquid water is stable in only a few places such as Earth's surface, and perhaps in underground oceans on some of Jupiter's several moons.

Given the right environment, however, 'chemical evolution occurs readily. Stanley Miller and Harold Urey demonstrated this in 1953 by creating a model of the early Earth in a lab flask. After only a few days relatively simple organic chemicals began forming complex amino acids, proteins involving combinations of some twenty amino acids being an essential part of all life forms (Sparrow 2006).

Thus, about 3.8 billion years ago Earth's watery environment became suitable for the development of life and prokaryotes developed in warm ponds or deep-sea hydrothermal vents, perhaps as both blue-green algae and bacteria (BBC 2014, Smithsonian Institute 2014).

Many believe that such life formed spontaneously in an appropriate 'chemical soup', but others view this as too improbable and believe that micro-organisms in meteorites from outer space began life on Earth, and that these gradually evolved into the myriad of life forms we see today.

Is There Life Out There?

As for the possibility of life elsewhere in the universe, Jayawardhana (2011) concludes,

It seems absurd, if not arrogant, to think that we are the only technological civilization in the Galaxy, given 200 billion other suns, the apparent ubiquity of planets, and the cosmic abundance of life's ingredients.

However it arrives, the first definite evidence of life—even primitive life—elsewhere will mark a revolution in science . . .

That dramatic moment is no longer a remote possibility; it may well occur in our lifetime, if not during the next decade.

Chapter 15
The Evolving Universe

> *The time has come, and indeed has long since come, to abandon the Big Bang as the primary model of cosmology. All the basic predictions of the Big Bang theory have been repeatedly refuted by observation. The theory is now cluttered with a multiplying collection of ad-hoc hypotheses.*
>
> Eric Lerner, Alternative Cosmology Group.

The Big Bang model is beyond repair

As noted in Chapter 9, there are now considerable grounds to doubt whether the universe is really expanding, the main basis for this belief being Hubble's law. To attribute redshift of radiation from distance galaxies to the Doppler effect, and thence to recession is clearly absurd, particularly because intergalactic light shows dispersion which is not taken into account in the Doppler effect.

Indeed, the redshift from some recently discovered quasars is so great as to imply a recession speed approaching that of light, whereas their brightness suggests that some of these quasars are not nearly as far away as their redshift implies. To deal with this issue, it has been proposed that quasars have some *intrinsic* component of redshift, yet another 'patch' to the increasingly ailing Big Bang theory.

In addition, beyond a boundary called the Hubble sphere, physics breaks down and objects recede at faster than the speed of light so that we can never detect them.

Finally, as noted in Chapter 9, angular size tests cast some doubt on the Big Bang theory, whilst within the redshift range 0.03 to 3.5 the static model provides a good fit with the data.

Cosmic dust

Hubble's law stating that the recession speed of distant galaxies increases with their distance is clearly absurd and, indeed, Hubble himself originally favoured the tired-light theory as an explanation for redshift.

Cosmic dust, however, is the most obvious explanation, the interstellar medium (ISM) containing 5% of the mass of the universe and about 10% of the mass of our galaxy according to Oxford Interactive Encyclopaedia (1997).

Most of the ISM is molecular atomic gas (99%), but cosmic dust plays a key role in star formation and mass loss when a star ends its life.

Interstellar dust ranges from large molecules such as polycyclic aromatic hydrocarbons (PAHs) to small carbon or silicate dust particles up to about half a millimetre in size, and PAHs dominate mid-infrared radiation (5-50 micrometres).

In our solar system a great deal of cosmic dust is found in the planetary rings that surround Saturn, Uranus, and Neptune, and also in the great many comets and meteors. Temperatures of dust particles in the asteroid belt are in the range 165-200 °K.

Near the sun, the dust particle density is 100 particles/cc, this reducing with the inverse square of the distance from the sun to about 5 particles/cc near the Earth. Nevertheless, according to one estimate, up to 40,000 tons of cosmic dust reaches the Earth's surface every year.

Most of the cosmic dust that falls on earth comes from meteoroids, is 40% porous, and ranges in diameter from 0.05 to 0.5 mm.

Chapter 15

Redshift by cosmic dust

The quantum electrodynamic (QED) theory of redshift considers a galaxy photon absorbed by a cosmic dust particle as confined within the dust particle geometry.

According to the QED theory of redshift outlined in Chapter 9, absorbed photons are re-emitted at wavelength $L_0 = 2Dn$, where n is the index of refraction.

For amorphous silicate n = 1.45, giving for dust particles with D = 0.25 microns a redshift of the Lyman-alpha line at 0.121 microns to a red line at 0.725 microns with redshift $Z = (L_0 - L)/L$ = 4.8. Such a redshift would be interpreted by Hubble's law as implying a recession velocity over 90% that of light, when, in fact, the cause may have simply been cosmic dust (Prevenslik 2010).

Thus refraction by cosmic dust seems the 'natural' explanation of cosmological redshift, whilst gravitational effects may also contribute to it.

Cosmic dust and the CMB

That there is a spectrum of background radiation is to be expected. As noted in the first section of Chapter 11, the sources of infrared, X-ray, and gamma-ray background radiation are now known, for example up to 70% of cosmic infrared background radiation being emitted when galactic dust particles re-emit absorbed UV and visible light as IR.

Similarly, carbonaceous interstellar dust is perhaps the most likely source of the CMB (Hoyle, 1983).

On a universe-wide scale the distribution of cosmic dust would be relatively uniform, accounting for the uniformity of the CMB.

Another, possible source of the CMB is the SZE, which is briefly discussed in Chapter 11, and this would easily account for the small bumps observed in the CMB.

Noel's attribution of the CMB to interchange of microwave quanta between hydrogen atoms in the interstellar medium is, perhaps, another possible mechanism by which the CMB could be created, hydrogen atoms being about a hundred times more numerous than dust particles in the interstellar medium (Noel 2012).

Photons

Photons are elementary particles, being the quantum of light and all other forms of electromagnetic radiation which allow matter and radiation to be in thermal equilibrium.

The particle nature of photons was established by observation of Compton scattering of single photons by electrons in 1923.

Photons are massless, have no electric charge, are stable, and have two possible polarization states.

According to Planck's law of radiation, the energy and momentum of a photon depend only on its frequency or, inversely, its wavelength W:

$$\text{Energy} = hc/W; \text{Momentum} = h/W$$

where h is Planck's constant.

If photons were not strictly massless they would not move at the speed of light in a vacuum, and very low limits on any possible photon mass have been determined. Photons inside superconductors do develop nonzero effective rest mass, however, so that electromagnetic forces inside superconductors are only short-range.

Light slows in transparent media such as air, water and glass, the ratio by which it is slowed being the refractive index of the media.

The speed of light is also affected by gravitational effects, resulting in gravitational redshift (GRS).

At the quantum level a vacuum may be filled with a limited number of continuously appearing and disappearing short-lived particle pairs such as quark-antiquark pairs, a mechanism that would explain the magnetization and polarization of the vacuum. In such a model the speed of light would be dependent on variations in the vacuum properties of space or time, fluctuations of photon propagation time being estimated to be of the order of 50 attoseconds per square metre of crossed vacuum (Urban et al. 2013).

In another study, polarization of the vacuum was modelled by virtual charged particle pairs acting as electric dipoles. In this model the impedance of the vacuum depends only on the sum of the square of the electric charges of particles and not on their masses. Such a model would also influence the propagation times of light slightly (Leuchs and Sanchez-Soto 2013).

The foregoing studies may help explain the slight differences in photon speeds discussed in Chapter 6 that were observed in light from distant objects in the universe.

Relativity

As discussed in Chapter 6, besides recent evidence that the speed of light in a vacuum may not be constant after all, there have been many counter-arguments against the special theory of relativity (STR).

Like many others, Buenker disputes the length contraction predicted by STR. Another author writes:

> *The author has demonstrated in several previous papers that the Lorentz contraction should not exist, and, therefore will not be found by SIM [Space Interferometry Mission]. If the residual is found, then special relativity will, for the first time, have passed a direct test of one of its most fundamental predictions. If the residual is not found, then special relativity will have to be abandoned completely* (Renshaw 2014).

The 1971 Hafele and Keating caesium clock experiment was discussed in Chapter 6. This was in modest agreement with the time dilation prediction of STR, and the time expansion prediction of GTR (as a result of reduced gravity), but later corrections to their results cast doubt over them. In 2005 the experiment was repeated with more accurate clocks by the UK's National Physical Laboratory over a shorter journey (London-Washington D.C. return), obtaining results within 4% of the relativity predictions.

Buenker's alternative space-time transformation (ALT) outlined in Chapter 6 is premised on "the principle of simultaneity of events", removing another widespread cause of discontent felt with STR (Beunker 2007).

His ALT eliminates the difficulty of *"two clocks can both be running slower than each other at the same time or that rods can both be smaller than each other (Einstein's symmetry principle)."*

He concludes:

> *Instead the ALT allows one to return to the ancient principle of rationality (and objectivity) of measurement (PRM), that is, that all observers must agree on the ratio of any two physical quantities of the same type. The PRM is the essential basis for introducing a rationalized set of units such as the mks or cgs systems. Experiments with clocks on airplanes, rockets, centrifuges and satellites (GPS technology) indicate strongly that measurement is not symmetric but instead rational, and especially in the case of the GPS, that events do occur simultaneously for all observers after taking into account the differences in the rates of the clocks used to make the respective measurements. The latter conclusion is perfectly consistent with the relativity principle, but the ALT also emphasizes that the units in which the various law of physics are expressed vary systematically from one system to another depending upon their state of motion and position in a gravitational field.*

Photons and relativity

STR predicts that mass changes with velocity according to:

$$m' = m/(1 - v^2/c^2)^{1/2}$$

In the case of photons, however, it would seem more sensible to attribute their lack of mass to their speed being the invariant 'c' in a vacuum, whereas photons do have a nonzero effective rest mass near absolute zero temperature. Indeed, for all matter, including fundamental particles, temperature is an important factor in the level of physical and chemical activity. In the case of photons, if not all matter, therefore, it could be argued that the foregoing equation should be m' = m(1 - v^2/c^2)$^{1/2}$.

This then takes the same form as the STR length contraction: L' = L(1 - v^2/c^2)$^{1/2}$.

Indeed, in the case of light some form of length contraction is deemed to occur, necessitating definition of 'zero-length':

Zero-length is of great importance in optical phenomena, because in Einstein's geometry any element of the track of a light-pulse is a vector of zero-length; so that if there were no definite zero-length a pulse of light would not know what track it ought to take (Eddington 1924).

The changing solar system

The first section of Chapter 2 briefly describes how our solar system was formed as a coalescence of dust and gases from an exploding star, presumably one much larger than our sun. The remarkable tale of the evolution of bacterial, plant and animal life on Earth is then discussed, concluding that man's interference in the global ecosystem is likely to have devastating effects ultimately, including mass extinction of a large proportion of plant and animal species.

Chapter 3 concludes by calculating how much the Earth-moon distance might be increasing according to Hubble's law, the result being only 3 cm per year.

There are, however, great changes taking place in our solar system. These include the following (Millennium Group 2014, Rosenblum, 2014):

[1] Changes in the interplanetary medium (IPM):
   ➤ More frequent coronal mass ejections from the sun.
   ➤ Two new populations of cosmic particles in the Van Allen radiation belts.

The more 'highly charged' IPM increases planetary interactions and also stimulates greater solar activity.

[2] Greater solar activity:
- Greater solar cycle activity in the sun, which became highly evident in the twenty-second cycle.
- More frequent class C flares.
- More frequent X-Ray Flux flares.
- The sun's magnetic field has increased by 230% since 1901.

[3] Planetary changes such as:
- growth of dark spots on Pluto.
- auroras on Saturn.
- polar shifts of the magnetically conjugate planets Uranus and Neptune.
- doubling of the magnetic field intensity on Jupiter.
- increasing thickness of the Martian atmosphere.
- development of an atmosphere on the moon.
- increasing brightness of most planets.
- overall volcanic activity has increased 500% since 1975. On earth volcanic activity has increased 400% since 1973, and natural disasters by 410% between 1963 and 1993.

Of particular note, the thickness of glowing plasma at the leading edge of heliosphere has increased from 10 AU to 100 AU, a massive change (Rosenblum 2014).

This suggests that the solar system is moving into a more highly charged region of space which excites the heliosphere's leading-edge plasma, causing more of it to form. Eventually, as this 'charging' process continues, the wavelengths that the sun emits will change, and this may change the basic nature of all matter in the solar system (Rosenblum 2014).

This change in the heliosphere may be in part a result of a change in the direction from which neutral interstellar atoms flow into the solar system. This change was found by comparing observations by NASA's Interstellar Boundary Explorer (IBEX) with observations by eleven other spacecraft between 1972 and 2011 (Frisch et al. 2013).

Chapter 15

The evolving universe

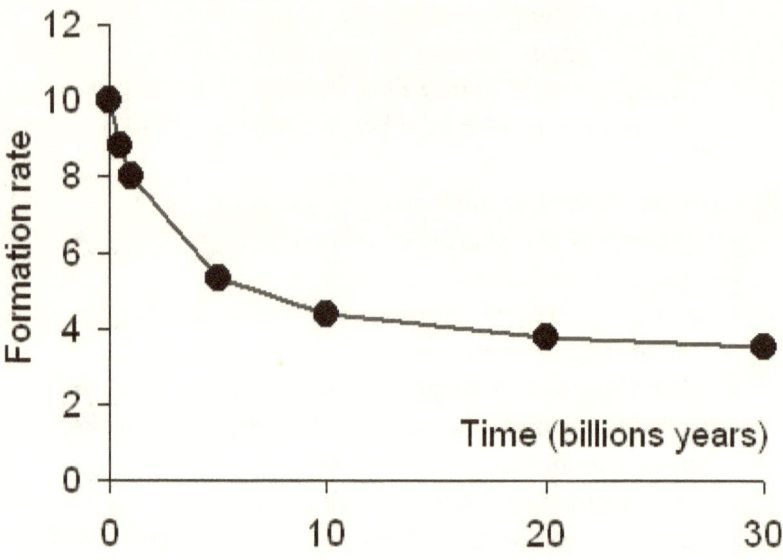

Figure 15.1. Star formation rate of the universe.

In like fashion to the many changes taking place in our solar system discussed in the preceding section, the universe is constantly changing, resulting in the formation of new planets, stars, and galaxies, along with the burning out of old stars (supernovas) and collapse of others to form black holes, some of which may take the form of quasars.

Astronomical records suggest that the rate of formation of new stars has decreased greatly over the known lifetime of the universe, and Figure 15.1 shows a hypothetical hyperbolic curve to illustrate this decrease. The equation for this curve is:

$$F = (3T + 25)/(T + 2.5); \; T \geq 0$$

where F is the 'formation rate', and T = time in billions of years. Thus F descends from an initial rate F = 10 at T = 0 towards the asymptote F = 3 at T = infinity, this being a final, evolved, stable state of the universe at which its entropy has reached some final limiting proportion of the total energy.

In the interim, F = 4.4 at T = 10, and F = 3.8 at T = 20, the current age of the universe according to the Big Bang theory being between these last two points.

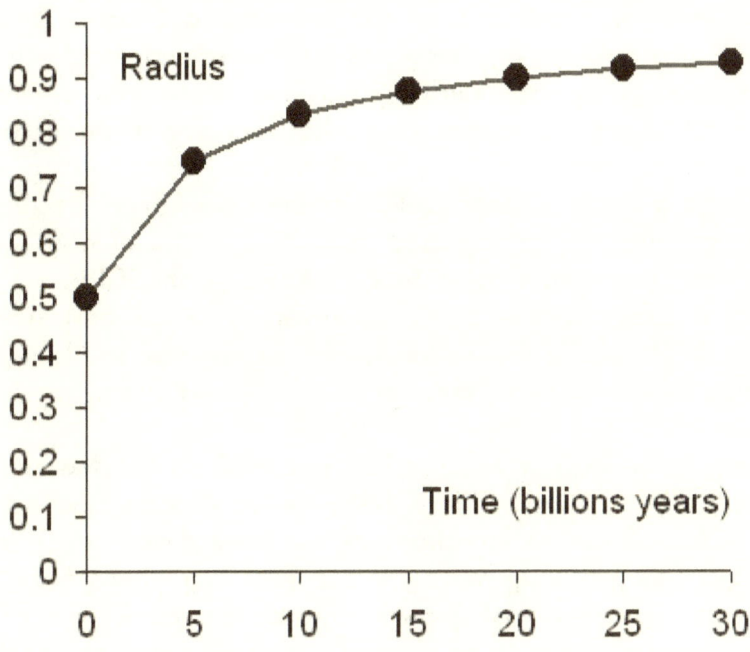

Figure 15.2. Slowing expansion of the universe.

If the universe is expanding and, as discussed in Chapter 9, there is much doubt about this, such expansion could take the hyperbolic form shown in Figure 15.2.

The equation for this curve is:

$$R = (T + 2.5)/(T + 5); T \geq 0$$

where R is the radius of the universe, and T is time in billions of years. Thus R increases from an initial radius R = 0.5 at T = 0 towards the asymptote R = 1 at T = infinity, this being a final, evolved, stable state of the universe.

Chapter 15

In the interim, R = 0.83 at T = 10, and R = 0.9 at T = 20, the current age of the universe according to the Big Bang theory being between these last two points.

Thus, as the rate of star etc. formation declines (Figure 15.1), so the rate of expansion of the universe decreases and the universe asymptotically approaches a fully evolved, stable state.

An analogy for this evolving universe might well be that of Earth itself where it certainly seems that evolution of an incredible number of life forms must have finally reached some limiting value. Indeed, as a result of degradation of the environment because of man's activities, we must expect a decline in the number of life forms from this point, if not a drastic decline (Mohr 2012).

Man's effect on the Earth, however, is in all probability a unique occurrence in the universe, and the universe itself might be expected to slow in evolution, not reverse with a diminution in size or content.

Note, however, that Figures 15.1 and 15.2 do not deal with the periods prior to T = 0. This is because we can only go back in time as far as our observations of the most distant stars allow, this being 13.7 billion years. Before that time there may have been a single period of growth towards the peak rate of star formation, or perhaps some prior slightly cyclic changes as in the quasi-steady state cosmology. About this earlier period, however, the best that any 'theory' of the universe can do is guess.

Star creation in the Milky Way

NASA studies using the European Space Agency's INTEGRAL satellite to study the distribution of radioactive aluminium-26 in our galaxy found that it produces an average of seven new stars per year.

Aluminium-26 is produced in supernova explosions and emits gamma ray radiation. NASA estimated that there are about three solar masses of it distributed fairly evenly throughout the Milky Way.

Star formation in the Milky Way is obscured by clouds of gas and dust in its spiral arms, but gamma rays can penetrate these clouds, providing the most direct means of observing stellar activity. Because aluminium-26 has a half-life of circa 750,000 years it enables scientists to trace the history of stars in our galaxy for the past several million years. Only a relatively small amount of aluminium-26 is created in a single star explosion, but theories of how much this amount is allow estimation of star death and formation rates.

There is a supernova event two or three times per century in the Milky Way, stars becoming supernovas being at least 10 solar masses.

Thus, one supernova event about every forty years equates to the loss of only circa 0.5 solar masses per year. Only a few stars end up as supernova, however, most ending as white dwarfs, later becoming brown dwarfs. Thus we must assume the Milky Way loses circa five solar masses per year, comparable to seven new stars with a total of about four solar masses per year estimated by NASA.

Thus, with the Milky Way now being about 10 billion years old and having converted about 90% of its initial gas content into stars, it seems to be reaching a substantial degree of equilibrium, as illustrated in Figure 15.1.

Similarly, it has been estimated that star formation in the local universe peaked about 11 billion years ago, and that it is now only about 3% of its peak value (Space.com 2012). This suggests a potential crisis in the future, but studies further afield in the universe suggest that star formation rates in larger galaxies are higher than in the past, so that the universe as a whole may indeed be following the slowdown towards equilibrium illustrated in Figure 15.1 (Bell et al. 2005, Kistler et al. 2013).

Conclusions

The Big Bang model is under increasing attack, cosmic dust, along perhaps with gravitational redshift, being the obvious explanation for the redshift that is attributed to recession velocity in the seemingly ridiculous Hubble's law, the more so because Hubble himself initially favoured tired-light theories.

The theory of relativity also seems under threat, predictions of 'reciprocal' observations of time dilation and length contraction being absurd. Buenker's alternative relativity model maintains simultaneity of events, solving one of the main objections to STR and GTR.

Chapter 15

Time dilation as a result of gravitation and relative velocity does seem to occur, however, between different frames of reference. As Buenker points out, this can be taken into account by allowing the units in which the various law of physics are expressed to vary from one system to another depending upon their state of motion and position in a gravitational field (Buenker 2007).

Then, as with Mohr's large curvature correction (Mohr 1981, 2001), there is no need to view the universe in some curved space-time. If a structural beam or a beam of light bends, that is best viewed and understood from a fixed orthogonal frame of reference as in Appendices B and C.

As for the mass increase with increasing velocity which the special theory of relativity predicts, the present authors feel that photons seem to behave oppositely and that perhaps this is yet another example of the failings of the theory of relativity.

Finally, the present authors favour the evolving universe model illustrated in Figures 15.1 and 15.2. In this the known decrease in rate of star and galaxy formation in the universe eventually plateaus and the expansion of the universe (if any, that being in great doubt as discussed in Chapter 9) ceases.

In the light of evolution as we know of it and see it on earth, the 'EU' model makes far more sense than such impossible concepts as the 'something from nothing' Big Bang and 'back to nothing' Big Crunch. Indeed, for the universe some eventually relatively stable, static form seems the most probable outcome, whereas individual solar systems are not permanent.

Chapter 16
Conclusions

> He gave man speech, and speech created thought,
> Which is the measure of the universe.
>
> Percy Bysshe Shelley, *Prometheus Unbound*.
>
> Whatever universe a professor believes in must at any rate be a universe that lends itself to lengthy discourse. A universe definable in two sentences is something for which the professorial intellect has no use. No faith in anything of that cheap kind!
>
> William James, *The Present Dilemma in Philosophy*.

Introduction

The two main theories of the universe in the last century were the steady-state theory and the Big Bang theory. Both were modified a good deal to explain new astronomical findings. The two main modifications of the steady-state theory were postulation of C-fields for creation of matter to maintain a constant density of the expanding universe, and the cyclic quasi-steady state cosmology (QSCC) which proposed that the universe was currently in an expansion phase to explain redshift and the CMB.

The main modification of the Big Bang theory was the proposal of an extremely short inflation period to help explain the uniformity of the CMB.

Chapter 16

There have, of course, been many other theories of the universe, the main 'scientific' theories having been discussed in Chapter 12, whilst religious views and 'alien' theories were discussed in Chapter 13.

Other theories of the universe have included the following:

- The universe had no beginning—proposed in the 1970s.
- The universe split into two 'parallel' universes—proposed circa 1994.
- The varying speed of light theory proposed circa 1998.
- The universe created itself—first paper in 1998.

The latter theory involved string theory, as did the relatively recent ekpyrotic theory which originated circa 2001 and is discussed in Chapter 12.

The theory of relativity

The theory of relativity contributed much to the development of the Big Bang theory, notions of curved space-time and thus a 'light cone' that extended both forward and backward in time providing a convenient way of estimating when the Big Bang occurred (Hawking 1989).

Eddington's simple proof of the special theory of relativity (STR) formula for time dilation is given in Chapter 6. This adds time to the three spatial dimensions as the 'imaginary' part of a complex number. This artifice is normally used to represent the amplitude of the usually sinusoidal variation of electromagnetic wave intensity with time, the 'real' part of the complex number representing the mean intensity.

Whether this complex number artifice is justified as a means of obtaining the time dilation formula is doubtful and, in fact, the fourth time dimension should be brought into play as a result of the coordinates (x,y,z) varying according to the components of velocity in these directions.

Conclusions

The STR formulas for length contraction and mass increase as a function of 'relative velocity' v/c are given without proof in Chapter 6, but the proofs of these formulas given by Eddington (1924) are also somewhat artificial.

Eddington's proof of Einstein's famous $E = mc^2$ formula, also given in Chapter 6, is quite 'loose' and more of an 'argument' than a proof.

Today, therefore, many scientists doubt the assumptions that led to the formula for the Lorentz-Fitzgerald length contraction which was the basis for Einstein's STR. In particular, many find objection to the STR prediction of 'time dilation' such that clocks in two frames of reference with different speeds can both be said to be running slower relative to the other.

An example of a 'dodgy' relativistic proof which leads to this paradox is given by Smid (2014). He begins with the standard Lorentz transformation for space and time:

$$x' = (x - vt)/F;\ t' = (t - vx/c^2)/F$$

where $F = \sqrt{1 - v^2/c^2}$ is the Lorentz factor and c is the velocity of light.

Einstein derived the formula for time dilation by setting the origin for the moving frame as $x' = 0$, so that it follows from the first formula above that $x = vt$. Inserting this result in the second formula above yields:

$$t' = \mathbf{F}\,t = \sqrt{1 - v^2/c^2}\,t$$

If, on the other hand, we consider the inverse Lorentz transformation we have:

$$x = (x' + vt')/F;\ t = (t' + vx'/c^2)/F$$

Now setting our origin as $x = 0$, we obtain from the first formula above $x' = -vt'$, and inserting this result in the second formula above yields:

$$t = \mathbf{F}\,t' = \sqrt{1 - v^2/c^2}\,t'$$

which is in complete contradiction of the first result $t' = \mathbf{F}\,t$.

Chapter 16

Similarly, putting t = 0 and t' = 0 respectively in the equations for x' and x above yields the same contradiction for length contraction, namely x' = x/F and x = x'/F.

Smid also gives a 'thought experiment' that illustrates the paradox of two space ships in relative motion by showing, on the basis of the usual relativistic time dilation theory, that an explosion occurring when they pass each other would only be observed in one of the two ships (Smid 2014).

Not surprisingly, therefore, as noted in Chapter 15 there is a growing feeling that the special theory of relativity should be abandoned.

One solution is the alternative Lorentz transformation (ALT) proposed by Beunker and discussed in Chapter 6, as this preserves simultaneity of events. Thus, as noted in Chapter 15, his ALT eliminates the difficulty of *"two clocks can both be running slower than each other at the same time or that rods can both be smaller than each other (Einstein's symmetry principle)"* (Buenker 2007).

Thus, whether a structural beam is curved, as with Mohr's large curvature correction (Mohr 1981, 2001), or a beam of light is curved, it is still best to view the universe or any 'structure' using an orthogonal frame of reference, keeping time as an absolute rather than 'relative' quantity.

That gravitational effects should sometimes be taken account of, however, is not in doubt, the 'tired-light' theory of redshift discussed in Chapter 9 being an example of this. Indeed, it is gravitational and inertial effects that must be deemed to be the reason why atomic clocks in satellites run slowly, and this should be seen not as relativistic time dilation, but more as the result of physical effects. After all, gravitational and inertial forces are amongst the most significant and pervasive in the physical world and, as shown in Appendix B, gravity can bend light and therefore might also be expected to affect the nuclear reactions in atomic clocks.

Conclusions

The Big Bang theory

At present, there is growing discontent with the Big Bang theory of the universe, particularly because it has had to be adjusted many times to account for new astronomical findings. The major adjustment was proposal of an incredibly short 'inflation' period in which the universe is supposed to have grown from an infinitesimal size to an almost infinite one.

As discussed in Chapter 8, this adjustment overcomes several problems, including the light-horizon, 'flatness' and 'monopole' problems.

The major difficulty of the Big Bang theory, of course, is its 'something from nothing' nature, and in that regard it has something in common with, and has been supported by, proponents of creationist arguments.

As a result of the many problems faced by the Big Bang theory various new theories have been proposed, varying from those involving string theory, which envision our universe as a 3-D brane in a 4-D bulk universe, to more ridiculous notions such as alien design.

The steady-state theory

The steady-state theory also underwent a few adjustments over time, principally the proposal of creation fields to keep the average density of the universe constant as it expanded.

The quasi-steady state cosmology (QSSC) attempted to explain the surplus of faint radio sources and the evolution of the CMB by proposing that the growth rate of the universe varied sinusoidally over time but with a diminishing amplitude of variation (Hoyle et al. 1993). The period of oscillations was assumed to be about three times the age of the Big Bang universe and the 'bursts' of creation in this theory are sometimes referred to as *little bangs*, it being proposed that quasar-like objects at the centre of galaxies might be responsible for these.

The cyclic nature of the QSSC provides an explanation for redshift, namely that the universe is currently in an expansion phase, whist the CMB is held to be caused by carbon and iron whiskers in the interstellar medium.

Chapter 16

Proponents of the QSSC argue that findings of many relatively young galaxies in a corner of our galaxy disprove the Big Bang concept of almost instantaneous creation of the universe.

The evolving universe theory

As with other cyclic theories of the universe, the QSSC assumes a time frame that in all likelihood will never be verified, and the authors believe the evolving universe model discussed in Chapter 15 is the 'best of both worlds', involving a declining rate of star etc. formation (Figure 15.1), so that the rate of expansion of the universe decreases (Figure 15.2) and the universe asymptotically approaches a fully evolved, stable state.

In the Milky Way, in which there is a good deal of interstellar dust and gas, it does appear that new stars are being formed at a rate comparable to that at which old stars disappear in supernova events or as white dwarfs.

In much of the universe star formation rates are only a small fraction (as little as about 5%) of what they were at their peak about 11 billion years ago. Recent findings suggest, however, that star formation rates are higher in more remote regions of the universe.

These findings do suggest that we are well down the hyperbolic curve of Figure 15.1 and, hopefully, on the way to a relatively stable and static universe.

Here it should be noted that our observations are able to go back in time only as far as the most distant objects we are able to observe imply, and beyond that any 'theories' as to 'what started it all?' or 'why does anything exist?' can only be guesswork.

Conclusion

Many alternative theories of the universe have been proposed over the last hundred years, the Big Bang theory having overtaken the steady-state theory in popularity in that time. The Big Bang theory, however, has been adjusted many times to account for new astronomical findings, and a new, more rational theory is needed that does not involve such problems as that of 'something from nothing' in the Big Bang, and the 'everything to nothing'

Conclusions

proposal of the Big Crunch which might one day appear in such books as *History's Worst Predictions* (Chaline 2011).

The theory of relativity did much to encourage the Big Bang theory of the universe, despite itself being fraught with often absurd predictions based on the constancy of the speed of light which could not be exceeded. Contradicting this, in Cherenkov radiation electrons from nuclear reactors travel through shielding water faster than the speed of light (in water). For several decades, however, objections to relativity theory were often suppressed but there is now increasing discontent with it and Buenker's alternative theory discussed in Chapter 6 takes the important step of ensuring simultaneity of events.

One of the main pillars of the Big Bang theory was the assumption of the expanding universe implicit in Hubble's law which assumes redshift is caused by the Doppler effect and thus absurdly accords objects recession speeds proportional to their distance from Earth.

Cosmic dust, however, perhaps in conjunction with gravitational effects, provides the most rational explanation of redshift of light from distant planets, whereas attribution in Hubble's Law of redshift to recession velocity that increases with distance from Earth is clearly absurd.

Cosmic dust also provides an adequate explanation for the cosmic microwave background radiation, whilst the Sunyaev-Zel'dovich effect and light interaction with interstellar hydrogen atoms may also contribute to the CMB.

Such explanations make the evolving universe the most rational and likely model of the universe, involving as it does change, but eventually reaching a relatively steady state as most physical and chemical processes do without external interference and given time.

As discussed in Chapter 10, the results of recent research confirm to a high degree of accuracy that the universe is flat and thus infinite. This in turn suggests that, as already noted above and discussed in Chapter 9, the universe is not expanding as much as the Doppler effect interpretation of redshift implies, such redshift being most naturally explained as being caused by cosmic dust.

Chapter 16

As for the origins of life on Earth, most scientists now believe that life on earth began about 3.8 billion years ago with prokaryotes in warm ponds or deep-sea hydrothermal vents, perhaps as both blue-green algae and bacteria, Wordweb 6 defining bacteria as

> *(microbiology) single-celled or noncellular spherical or spiral or rod-shaped organisms lacking chlorophyll that reproduce by fission; important as pathogens and for biochemical properties; taxonomy is difficult; often considered to be plants.*

As discussed in Chapter 14, however, there has been much evidence of bacterial fossils in meteorites found on Earth, suggesting that life has developed elsewhere in the solar system (Jayawardhana 2011). Indeed, Hoyle argues that, not only were bacteria from outer space the probable origin of life on Earth, but also that viruses from outer space might also have contributed to the evolution of life on Earth, and that they might still be responsible for epidemics of such diseases as influenza (Hoyle 1983).

This latter possibility is unlikely, however, but more likely is that the first life on Earth might have been dormant bacteria aboard fragments of a meteor from Mars (for example only) which landed in a warm pond or hot ocean vent, and these bacteria could have played host to viruses.

Indeed, as discussed in Chapter 14, the possibility of the appropriate DNA sequences developing randomly in only the first few hundred thousand years after which the Earth's environment ceased being too hostile for life is perhaps remote, favouring the 'life from space' theory.

The evolution of life on Earth is as outlined in Chapter 2 and summarized in Appendix A, the enormous multiplicity of life forms now evident probably having peaked. Indeed, in part as a result of mankind's excessive population, and its consumptive and destructive effects on the planet, we should expect a considerable reduction in animal and plant species on the planet over the next millennium, and perhaps a considerable reduction in our own numbers too as a result of competition for fast diminishing

Conclusions

resources and rapidly evolving viruses (Palumbi 2001, Cullen 2010, Mohr, 2012).

As for the universe, star formation rates have declined greatly from their peak billions of years ago, but the rates in our galaxy, and more distant galaxies in the universe, seem of a sufficient magnitude to maintain the sort of long-term equilibrium illustrated for the evolving universe theory in Figure 15.1, at least for the foreseeable future.

Certainly a settled state of equilibrium is what we should hope for at all levels ranging from those of human affairs to those of the cosmos.

Atoms, which in the Bohr-Rutherford model are like a miniature solar system, are stable. Our solar system is stable. So too should the universe be able to ultimately achieve a stable or steady state. This is the basic tenet of the evolving universe theory.

Finally, it should be noted that at present our understanding of the universe is still only relatively primitive, leaving many questions requiring further consideration, some of these being summarized in Appendix E.

Appendix A
Earth's Evolutionary Timeline

The Hadean eon

This first eon began with the formation of the Earth about 4.6 billion (4,600 million) years ago from the accretion disk revolving around the sun. There is no evidence of life in this period but complex organic molecules may have formed in the cosmic dust surrounding the sun before the Earth was formed.

According to the 'giant impact hypothesis' the moon was formed circa 4.5 billion years ago when the Earth and planet Theia collided, producing many 'moonlets' orbiting Earth which eventually combined to form the moon. Then the gravitational effects of the moon stabilized the Earth's shifting axis of rotation, improving its prospects of eventually developing life.

The period 4.1 to 3.8 billion years ago was that of the Late Heavy Bombardment (LHB) during which a barrage of meteors impacted the inner planets.

Earth's oldest known rocks date back about four billion years to the beginning of the Archean eon.

The Archean eon

This extended from about 4 billion years ago until 2.5 billion years ago, during which time the earth's crust formed and life began to evolve.

Appendix A

Conditions were hostile to life until the end of the LHB circa 3.8 billion years ago, when thermal flux from widespread hydrothermal activity may have been conducive to the evolution of the first simple cells, prokaryotes, primitive organisms without a cellular nucleus, these mainly being bacteria. These were chemoautotrophs which use carbon dioxide as a carbon source and oxidize inorganic materials to obtain energy.

About 3.5 billion years ago a second basic type of living organism, archaea, evolved to use glycolysis, chemical reactions that obtain energy from organic molecules such as glucose and store it in the chemical bonds of adenosine triphosphate (ATP). This nucleotide is stored in muscle tissue and is the major source of energy for cellular reactions in almost all organisms.

The first archaea lived in deep-sea hydrothermal vents using a process known as chemosynthesis to obtain energy from chemical reactions involving hydrogen sulphide and other inorganic compounds. These deep-sea archaea are the bottom of the food chain for clams, tube worms, mussels, and other animals that live near the vents. Archaea have also been found in the guts of animals, compost piles, saturated marshes, and other common places.

The earliest evidence of life on Earth is biogenic graphite in 3.7 billion year-old meta-sedimentary rocks in Western Greenland, and 3.5 billion year-old microbial mat fossils found in sandstone in Western Australia.

Circa 3 billion years ago cyanobacteria evolved using water as a reducing agent and producing oxygen as a waste product. In this period the moon was still very close to the Earth, causing 300 metre tides accompanied by hurricanes, these conditions being thought responsible for the first land-based life forms.

Circa 2 billion years ago, eukaryotes evolved as organisms with a cellular nucleus containing chromosomes, these being fungi, plants, and animals.

A chain of evolution

Earth's first form of life was that of prokaryotes about 3.7 billion years ago, beginning the 'chain of evolution' shown in the Table A.1.

Earth's Evolutionary Timeline

Table A.1. Chronology of evolution of life on Earth.

Period	Life form
Last 3.7 B years	Prokaryotes
Last 3.5 B years	Archaea
Last 3 B years	Cyanobacteria
Last 2 B years	Eukaryotes
Last 1 B years	Multicellular life forms
Last 600 M years	Simple animals
Last 550 M years	Animals with a front and a back (bilaterians)
Last 500 M years	Fish and proto-amphibians
Last 475 M years	Land plants
Last 400 M years	Insects and seeds
Last 360 M years	Amphibians
Last 300 M years	Reptiles
Last 200 M years	Mammals
Last 150 M years	Birds
Last 130 M years	Flowers
Last 60 M years	Primates
Last 20 M years	Hominidae (great apes)
Last 2 M years	The Homo genus
Last 250,000 years	Modern humans (homo sapiens)

As Table A.1 clearly shows, the Homo genus, the forerunner of modern man, appeared only relatively recently, and the evolution of modern man, or *Homo sapiens sapiens*, is discussed in Chapter 2.

As discussed in Chapter 14, some scientists hold that there was not sufficient time after the LHB for the first life on Earth (prokaryotes) to evolve by gradual changes in DNA, and that it is more likely that these first bacteria were transported to Earth in cosmic dust or aboard meteorites.

Appendix B
Gravitational Deflection of Light

Introduction

The foundation of the special theory of relativity (STR) is the assumed invariance of the speed of light, and STR was briefly discussed in Chapter 6, including a simple proof of 'time dilation' using complex numbers. Buenker's alternative approach to relativity which preserves simultaneity of events was also briefly described.

The universe is governed by four fundamental forces, the short-range strong and weak forces at subatomic level, the electromagnetic force (emf), and gravitational force. The emf is a 'chain' type of effect, electromagnetic radiation involving a 'knock-on' effect in which neighbouring particles propagate the emf.

Gravitional forces, of course, are long-range ones and the purpose of Einstein's general theory of relativity (GTR) was to include these. As noted in Chapter 6, the classical prediction of GTR was gravitational deflection of light passing the sun, the results of numerous measurements of this deflection being given in Chapter 6, the average finding being a deflection of about 2 seconds of arc.

Gravitational Deflection of Light

In the following section a simple calculation that has no need of recourse to the confusing complexities of tensor calculus is given (tensors being 'generalized vectors').

Gravitational deflection of light

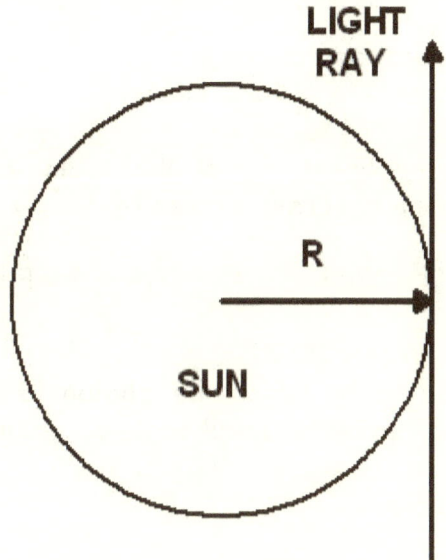

Figure B.1. Passage of a light ray past the sun.

Figure B.1 shows a light ray passing close to the sun. Here the gravitational pull of the sun gives a transverse velocity U calculated by integrating the acceleration given by the fundamental relationship F = ma, giving

$$U = \text{Integral}[a\, dT] = \text{Integral}[(F_x/m_p)dT]$$

where a is the acceleration of a photon, F_x is the lateral force of gravity on the photon, m_p is the mass of the photon, and T is time. Using Newton's aw of gravitation the radial force of gravity F_G is given by

$$F_G = G(m_s m_p / r^2)$$

Appendix B

where G is the gravitational constant, m_s is the mass of the sun which is taken to be concentrated at its centre (Einstein 1922), and r is the distance of the photon from the centre of the sun which is given by

$$r = (R^2 + Y^2)^{1/2} = R[1 + (Y/R)^2]^{1/2}$$

where Y is the vertical coordinate in Figure B.1 measured from the point of closest contact with the sun, at which point r = R, where R is the sun's radius.

The lateral component, F_x, of the radial gravity force F_G (perpendicular to the light ray) is given by

$$F_x = F_G R/(R^2 + Y^2)^{1/2} = F_G/[1 + (Y/R)^2]^{1/2} = F_G/F$$

where F is the factor $[1 + (Y/R)]^{1/2}$.

Finally, the angle by which the photon in the light ray is deflected is then given by the small angle approximation

Angle = U/c (in radians)

Combining these results gives

Angle = Integral$[(F_G/m_p Fc)dT]$
      = Integral$[Gm_s m_p/r^2 m_p Fc)dT]$
      = $(Gm_s/R^2 c)$ Integral$[dT/F^3]$

and the photon mass cancels out of calculations.

Taking time to be a function of change in the coordinate Y in relation to the speed of light one writes dT = dY/c, and taking Y = sR where s is a dimensionless coordinate, this transformation becomes dT = (R/c)ds, so that the final calculation for the light angle is

$$\text{Angle} = (Gm_s/Rc^2) \text{ Integral}[ds/F^3]$$

where $F = (1 + s^2)^{1/2}$.

Gravitational Deflection of Light

Integration of the term (ds/F^3) can then be carried out numerically as

$$\text{Integral} = \text{Sum}[1/(1 + s^2)^{3/2}] \times \text{step length}$$

Here the required integral is calculated at each of several points equally spaced by a 'step length' L, coding for this calculation in BASIC being

```
S = 0: L = 1
FOR N = -100 to 100
Y = N * L: P = 1/(1 + Y^2)^1.5
S = S + P * L: NEXT: PRINT S
```

The results with various values of step length L are:

L	0.3	0.4	0.5	0.6	0.8	1	1.2
S	1.9989	1.9994	1.9997	2.0002	2.0056	2.0248	2.0656

These values show good 'stability' with a step size of about 0.5, giving a result of about 2 for the integral. This corresponds to a physical step size of 0.5R, about the right magnitude.

Now deflection of the transverse component of the light ray must be considered. This is only a proportion U/c of the original light ray, but travels far more slowly, so that integration of time must be factored by the inverse of U, not the large value c. The two factors here thus cancel to unity, so that adding the deflection of the transverse light component doubles the total deflection of the light ray so that, including the result 2 for the integral term obtained above, the total deflection of the light ray is

$$4Gm_s/Rc^2$$

which is the same result as that obtained by Einstein (Einstein 1922). Substituting the known (rounded) values in the factor (Gm_s/Rc^2) in mks units gives this as

$$(Gm_s/Rc^2) = (6.67 \times 10^{-11})(2 \times 10^{30})/[(0.7 \times 10^9)(3 \times 10^8)^2]$$
$$= 2.12 \times 10^{-6}$$

giving the predicted angle as $4 \times 2.12 \times 10^{-6}$ radians, or $8.48 \times 10^{-6} \times 206265 = 1.75$ arcseconds.

As discussed in Chapter 6, a number of observations over several decades were in general agreement with this prediction.

Note that in the foregoing calculation the simple doubling when deflection of the slower transverse light component is taken into account avoids the need to consider curvature in space-time. Note also that Einstein's assumption of the sun's mass being concentrated at its centre was a 'significant' approximation.

Propagation of gravity

Newton regarded notions that gravity's effects were somehow propagated instantly through an infinite 'nothingness' as absurd. The general theory of relativity predicts gravitational waves that move at the speed of light, such waves being considered to be 'ripples' in the curved space-time continuum.

Such 'ripples' are caused by a sudden and significantly large change in a gravitational field such as removal of a large mass.

Observations of the orbital decay of binary pulsars PSR 1913+16 and PSR B1534+12 as a result of lessening gravitational radiation, or 'gravitational damping', indicated that the speed of gravity was within 1% of the speed of light. That such gravitational damping occurs at all implies that the speed of gravity is finite.

In November 2013, however, Y. Zhu announced that he had calculated the speed of the gravitational force by observing variations in the orbit of geosynchronous satellites, finding that it was much greater than the speed of light in a vacuum.

In a 1998 paper published in the journal *Physics Letters*, the speed of gravity was calculated as greater than $2 \times 10^{20}c$, an enormous but finite speed (van Flandern 1998).

Hopefully research will soon establish an undisputed speed of gravity. As for the 'medium' by which gravity is propagated, this may turn out to be the as yet undetected graviton.

The three other fundamental forces are mediated by elementary particles with spin-1: electromagnetism by photons, the strong force by gluons, and the weak force by the W and Z bosons, whereas 'matter' is made up of quarks and leptons with spin-1/2. The newly discovered Higgs boson has spin-0, and some scientists believe that the graviton must be a massless spin-2 boson.

According to the 'electric universe' theory electromagnetic radiation is propagated as transverse polarisation of an 'aether' of neutrinos, whereas gravitational radiation is a longitudinal polarisation of these particles (Sykes 2012).

Dannon (2013) also proposes that it is neutrinos that propagate gravitation, arguing that gravitational radiation is emitted when a quark jumps from a down-quark energy level to an up-quark energy level with the emission of an electron and a neutrino = graviton. This transition would convert a neutron (2 d-quarks and 1 u-quark) to a proton (1 d-quark and 2 u-quarks).

Certainly neutrinos are all-pervasive in the universe, thousands of them passing through our bodies every second without interaction. These elusive particles can even pass through the entire Earth without being affected. They were postulated in 1931 by Pauli to account for an apparent loss of energy in beta decay, and finally detected in a study of particles created in a nuclear power plant.

The sun emits neutrinos by combining a pair of protons to form a deuterium nucleus with the emission of a positron and a neutrino whilst collapsing supernova cores combine protons and electrons to form a neutron and a neutrino.

Neutrinos are fermions with spin-1/2, however, not spin-2 bosons as theory has predicted that gravitons should be, so more work needs to be done to confirm whether neutrinos might be the medium for gravity, or whether gravitons do actually exist and carry gravity. Another difficulty is that dark matter, postulated as one of the main causes of gravitational effects in the universe, has itself not yet been proven to exist.

Appendix C

The Large Curvature Correction In FEM, by G. A. Mohr

Introduction

The LCC for finite element analysis (FEM) was first proposed in Cartesian form by Mohr in 1979 (Mohr and Milner 1981), and Milner added a curvilinear form for this correction (Milner 1981). The original Cartesian correction had a slight error which did not affect results noticeably, whilst the original curvilinear correction claimed impossibly good results with only one or two elements and one load step, thanks to an erroneous, if not ridiculous, cosine term associated with moment loadings (Mohr and Argyris 2001).

The work was corrected in a paper faxed to the late John Argyris in Stuttgart. On the phone, he applauded the work and I agreed to add to it and include his name. To that end I wrote a part II of the paper and posted this to Stuttgart. In the end, part I was lost whilst part II was typeset as CMA2085 and queued for printing in *Computer Methods and Applied Mechanics*, for which Argyris was chief editor.

When I complained to a US editor (JT Oden) about slowness reviewing a paper combining the part I and part II papers he gave me four insane telephone rants which I recorded (and kept) and the typeset part II stayed in the publication queue for a few years at Elsevier Science and then disappeared despite many enquiries about it to Nick Pinfield (PhD Camb.).

Circa 2006, I used many more elements to obtain very accurate results that did, at last, prove the large curvature correction (LCC) beyond all doubt, putting these results in a final combined paper. This paper was sent to Emma Hunt at the *Australian Maths Society* journal at Adelaide University but also ended up buried and never published, despite numerous enquiries and complaints to a female editor at Melbourne University (Dr A Tsordesillas).

To ameliorate that 'ungood' outcome the key equations and results from that work are given in this appendix.

The Cartesian LCC

Normally direct strains S_x in the x direction in a material are calculated as $S_x = du/dx$, where u, v are the Cartesian displacements in the x and y directions, and du/dx here denotes the partial derivative of u with respect to variation in the x direction only. If, for example, the *interpolation function* for u is the bilinear function $u = a + bx + cy$, then $du/dx = b$, and $du/dy = c$.

In a beam subject to bending the Cartesian curvature for small displacements is calculated as simply:

$$C = d^2v/dx^2$$

When the beam undergoes large displacements, however, the axial direct strain must be calculated as:

$$S_x = du/dx + (dv/dx)^2/2$$

Large displacements alter the geometry of the system and render the problem nonlinear so that a stepwise solution method must be used. For this a 'virtual' variation in the extensional strain is calculated as:

$$V(S_x) = V(du/dx) + V[(dv/dx)^2/2] = V(du/dx) + (dv/dx)V(dv/dx)$$

Appendix C

and this term is used to calculate *residual loads* which are recalculated at each step of the nonlinear solution procedure, and which gradually diminish as the accuracy of the solution improves.

For consistency, I decided to introduce the large curvature correction which in Cartesian form calculates the beam's curvature from the well-known formula:

$$C = d^2v/dx^2/F^3 \text{ where } F = [1 + (dv/dx)^2]^{1/2}$$

Then a virtual curvature increment is calculated as

$$\begin{aligned}V(C) &= V(d^2v/dx^2)F^{-3} + (d^2v/dx^2)\,V(F^{-3}) \\ &= V(d^2v/dx^2)F^{-3} - 3(d^2v/dx^2)(dv/dx)F^{-5}\,V(dv/dx) \\ &= V(d^2v/dx^2)[1 - 3(dv/dx)^2/F^2]/F^3\end{aligned}$$

here using reciprocity to exchange incrementation in dv/dx with incrementation in d^2v/dx^2 in the second term, and this LCC is included in the residual load calculation.

The curvilinear curvature correction

Alternatively, the beam's curvature can be calculated with respect to its curvilinear coordinate s, which follows the curved shape of the beam:

$$C = (d^2v/ds^2)/F \text{ where } F = [1 - (dv/ds)^2]^{1/2}$$

Then a virtual increment in this is given by

$$\begin{aligned}V(C) &= V(d^2v/ds^2)F^{-1} + (d^2v/ds^2)\,V(F^{-1}) \\ &= V(d^2v/ds^2)F^{-1} + (d^2v/ds^2)(dv/ds)F^{-3}\,V(dv/ds) \\ &= [V(d^2v/ds^2)/F][1 + (dv/ds)^2/F^2]\end{aligned}$$

and again reciprocity is used to obtain the final result.

Numerical results

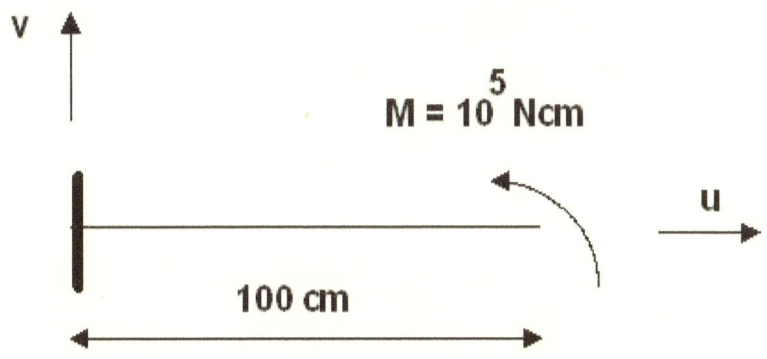

Figure. C.1. Large displacement of a beam,
$A = 10$ cm^2, $I = 5$ cm^4, $E = 2 \times 10^6$ cm^2

With the usual 3 degrees of freedom per node, namely u, v and dv/dx, the results obtained for the simple beam problem shown in Figure C.1 are shown in Table C.1.

Here, A = cross sectional area, I = moment of inertia of the beam cross section, and E = Young's modulus (of elasticity). For the FEM analysis 'e' = 4 or 10 elements are used with the total moment loading (M) at the end of the beam applied in 's' steps with 'i' iterations of the residual load solution technique used at each load step.

Here both the Cartesian and curvilinear LCCs require forty load steps to give a reasonably accurate solution for the lateral displacement 'v' at the end of the beam.

Appendix C

Table C.1. Results using 3 d.f./node.

e	s	i	u	v	dv/dx
With Cartesian curvature correction					
4	10	3	-9.3152	36.6826	0.7441
10	10	3	-13.7880	43.3140	0.9261
10	10	5	-13.7831	43.3070	0.9259
10	20	3	-15.2040	45.2057	0.9780
10	40	3	-15.6558	45.7853	0.9942
With curvilinear curvature correction					
4	10	3	-9.2879	36.6699	0.7401
10	10	3	-13.5845	43.1156	0.9125
10	10	5	-13.5724	43.0090	0.9121
10	20	3	-15.1177	45.1203	0.9730
10	40	3	-15.6305	45.7586	0.9928
Exact			-15.8529	45.9698	1.0000
Small displacement solution			0.0	50.0	1.0

As shown in Table C.2, better results are obtained using four freedoms/node, the added freedom being dv/dx.

Stress calculations for the problem are satisfactory (Mohr and Argyris 2001) and to prove the LCC beyond all doubt results were obtained using up to 100 elements and 100 load steps, as shown in Table C.3.

Table C.2. Results with 4 df/node.

e	s	i	u	v	dv/dx
With Cartesian LCC (4 df/node):					
20	10	3	-16.1959	46.3145	1.0173
20	20	3	-15.9300	46.0581	1.0043
20	40	3	-15.8656	45.9953	1.0010
With curvilinear LCC (4 df/node):					
20	10	3	-15.7576	45.8890	0.9934
20	20	3	-15.8224	45.9528	0.9983
20	40	3	-15.8388	45.9690	0.9996
Exact:			-15.8529	45.9698	1.0000

Table C.3. Tip deflection v using 3 iterations per load step

Load steps, s	e = 25	e = 50	e = 100
10	45.88721	45.88530	45.88483
20	45.95118	45.94906	45.94853
40	45.96736	45.96514	45.96436
50	45.96931	45.96708	45.96652
100	45.97193	45.96967	45.96910
Extrapolated	45.97270	45.97046	45.96984
Exact		45.96977	

Previous curvature correction results were obtained using single precision MegaBasic computation (8 figures). To provide the greater memory needed for larger problems Visual Basic was used to obtain the results of Table C.3. Single precision (7 figures) gave slight errors for e = 50 and serious errors for e = 100 so that double precision (15 figures) was used for all the results of Table C.3.

Appendix C

The results for increasing numbers of load steps are extrapolated by assuming that the solution error decreases according to the inverse square of the number of load steps so that the ordinal intercept of the regression line for the plot of v against $1/s^2$ is the extrapolation result. For all three columns of data a coefficient of correlation = 1 to 5 decimal places was obtained. Accurate regression lines for increasing e can be obtained in the same way.

It has been suggested to me that the foregoing large curvature correction might have application in the theory of relativity, but I doubt it. More obviously, small curvatures in light deflected by stars could be calculated using the small curvature approximation of structural mechanics, i.e. curvature = d^2v/dx^2.

The LCC does illustrate, however, that physical systems involving large curvatures can be studied with reference to an orthogonal frame of reference and there is no need to consider a curved coordinate system as is used in the general theory of relativity.

Appendix D

The Ridiculous Theory of Relativity

Introduction

The special theory of relativity was discussed at some length in Chapter 6, where some of the major misconceptions and errors involved in its derivation were pointed out in passing. As noted in that chapter, and others (particularly Chapter 16), many scientists and lay people today question the theory of relativity, especially such predictions as a clock being able to run slower in a moving frame of reference relative to another in a stationary one, and yet the reverse situation being able to apply also.

It was for this reason that Buenker's alternative theory of relativity was discussed briefly, because this ensures simultaneity of events. When the normalization factor (N) for this = 1, as will normally be the case, Buenker's alternative theory reduces to the Lorentz transformation, however, taking us back to 'square one' somewhat.

The purpose of this appendix, therefore, is to propose, somewhat tongue in cheek, the ridiculous theory of relativity (RTR), trying to avoid some of the same errors in both thinking and simple maths made by Einstein et al. circa a century ago, resulting in a hoax comparable in magnitude to those involved in launching most of the world's religions for political or monetary gain over the last couple of thousand years (Mohr and Fear 2014).

Appendix D

A brief history of the RTR

The RTR was developed in March 2014 by the first author, in consultation with a number of other people, including the other two authors of this book.[1]

It's starting point was the realization that there must, in fact, be some sort of 'aether' in space, despite the Michelson-Morley experiment failing to find that the speed of light was affected by direction of travel in space, leading to the conclusion at that time (1887) that there was no aether.

Now, however, dim-witted scientists, believing in the absurd Hubble's law, and thence that objects in the universe are receding away from Earth at great speed proportional to their distance from us,[2] find it necessary to postulate the existence of dark matter and dark energy to explain gravitational effects and accelerating expansion of the universe in the last few billion years.

The Higgs boson having been tentatively detected, it remains to detect the supposed carrier of gravity, the graviton, some believing that neutrinos may, in fact, play the role of transmitting gravity, as discussed in earlier chapters.

This then takes us full circle. Already we know that the interstellar medium contains massive amounts of gas and solid particles, as discussed in Chapter 5, and many believe that it is interstellar material that causes redshift[3], and also the cosmic microwave background[4], these questions being discussed in Chapters 9, 11, and 12.

It was the null result of the Michelson-Morley experiment that lead to the conclusion that the speed of light was constant, and thence to Fitzgerald's proposal of relativistic length contraction

---

[1] The author also consulted with his late father CBO, one of the world's leading theoretical physicists, about some key points that arise in the RTR.

[2] This mistaken belief arising because of attribution of redshift to the Doppler effect - see Chapter 9 for discussion of this issue.

[3] Not the Doppler effect.

[4] Not a result of the Big Bang.

with increasing velocity, and the Lorentz transformation that was the foundation stone of the special theory of relativity (STR).

It seems, therefore, that we are back to square one in finding it necessary to postulate some sort of particle by which gravity can be transmitted, namely gravitons which presumably play a role comparable to that of photons in the transmission of electromagnetic radiation. In addition, dark matter and dark energy have had to be proposed to prop up scientific 'weak spots' and ignorance. In the following section, therefore, the RTR is proposed as eleven laws.

The eleven laws of the RTR

The eleven fundamental laws of the RTR are shown in Table D.1, most of these having been verified by observation or accepted by general consensus. In the spirit of the 'baker's dozen' we have generously given an extra law over and above the number in the Ten Commandments supposedly delivered to a solitary and deluded Moses on stones atop a mountain of bullshit. The first is that the speed of light is not constant, in stark contrast to the key assumption of Einstein's STR. Thus, for example, just as light can be deflected by gravity (as simply demonstrated in Appendix B), so too its speed must be influenced by gravity (giving the second RTR law), so that light emerging from the most massive stars will be slowed, however slightly, this resulting in gravitational redshift (GRS).

In law 3, space is kept as 3-D, it being erroneous, if not downright stupid, to add time as a fourth dimension and thence 'fiddle' this fourth dimension by writing it as the imaginary part of a complex number and including in the term for this the speed of light (i.e., in Chapter 6, putting the T with spatial dimensions equal to icdt).

Appendix D

Table D.1. The eleven fundamental laws of the RTR.

Law	Statement
1	The speed of light (c) is not constant/invariable
2	c is not independent of it's source, e.g. GRS.
3	Space is 3-D, coordinates x,y,z being functions of velocities and thence time (if required)
4	A spatial dimension cannot be = i x c x dt
5	Time is absolute (ensuring simultaneity of events)
6	A 12" ruler is just that in any frame of reference
7	Mass is absolute and = rest mass
8	ds^2 does not equal $-ds^2$
9	You should not put the wrong signs in the Lorentz transformation
10	c is not = 1 (and then = c again)
11	E is not = mc^2, but E = BS^2

The fourth law disallows that latter ridiculous mathematical step.

In law 5, time is deemed absolute to ensure simultaneity of events. In the STR time differences between two frames with relative velocity arise because of the time light takes to travel between the frames, this a result of assuming this time difference being proportional to ds/c.

The sixth law denies the possibility of relativistic length contraction which has never been proven by observation. Indeed, the STR use of the Lorentz factor to predict that length becomes infinitesimal when an object reaches the speed of light is absurd in the extreme.

The seventh law makes mass absolute, rest mass being the real mass, of course, as noted in quoting Einstein on this point in Chapter 6. As he opined, any additional apparent mass is, indeed, in the form of energy.

Sure, nuclear reactions involve conversion of mass to energy, but subatomic physics is a realm in which most normal physical laws are apt to break down, including any that might be proposed in the STR or the new RTR which supersedes it.

The eighth law ensures that we can't change the sign of a mathematical quantity at whim, as does Eddington (1924, p22) in his proof of a relativistic velocity transformation.

The ninth law arises from Eddington having the equations for the Lorentz transformation with the wrong signs within them (they are correct in the second section of Chapter 6). A Cambridge man he was, of course, allowed to be an idiot, as was Newton and those that followed him, for example Hawking who derides Newton as being a rather nasty fellow (Hawking 1989).

The tenth law also comes from Eddington's BS (1924, 30) where he arbitrarily decides to put c = 1 for a while in order to struggle to yet another rather 'rigged' proof, this time leading us to the absurd m' = m/F where F is the Lorentz factor = sqrt(1 - v^2/c^2).

The same silly step is taken again on a temporary loss of sanity basis on page 32 in order to arrive at the famous E = mc^2 (Eddington 1924).

The final eleventh law is that E does not = mc^2. Proofs of this law are loose wherever you look, for example Eddington (1924) and Prokhovnik (1967). Prokhovnik does give modest discussion to the question of whether time should be absolute, but without reaching any conclusion as the present authors have in discussing Buenker's alternative Lorentz transformation in Chapter 6 (Buenker 2007).

Nuclear reactions may involve destruction of mass with the emission of energy, often as photons, just as smoke and light come from fires, but generally it is not possible to exactly determine some total energy of a body or particle and exactly equate this to its rest mass and its kinetic, heat etc. energy.

As for the farcical miscellany of misconceptions and errors involved in deriving the STR, the authors refer the reader to Mohr's ninth law, which is that Murphy is the true prophet (Mohr and Fear 2014). Mohr's tenth Law is that generally judgements should not be 'black and white', but out of 10, and we hope that readers will not judge this book too poorly.

Hopefully, they might, for example, give the simple proof of light deflection by the sun given in Appendix B at least a pass score, even though the calculations are crude at one point, involving a 'doubling' to account for the deflection of the (slower) transverse component of light's velocity that is created by the sun's gravity.

Appendix D

Finally, an addendum to the eleventh RTR law is Mohr's law of bullshit, $E = BS^2$ (Mohr 2012), where E is the effectiveness of the bullshit, B is the quantity of the bullshit, and S is its speed (i.e., number of repetitions per month or year). As in Einstein's law, we do not have acceleration from rest so the usual factor of 1/2 that is used to calculate kinetic energy is omitted.

This great law has always governed our fate, for example, the centuries of daily or weekly bullshit from religious leaders some of whom still encourage us to breed like animals, or the deluge of ads we *consumer zombies* now endure in our deranged consumer society (Mohr 2012). Those who produce this constant hail of BS should be locked up in a zoo or funny farm/giggle palace like Bedlam to do their preaching. The same applies to scientists and looney astrologists/astronomers who all too often espouse harebrained ideas such as those of the STR, the present RTR being intended as something of an antidote for Einstein's fantasies.

Appendix E

Many Remaining Questions

Many of the issues discussed in the foregoing book involve unanswered questions. Just as gods are hypothesized in religion to explain how the universe and life on Earth came to be, so modern science has increasingly postulated such unlikely events as the 'something from nothing' Big Bang, and as yet undetected entities such as dark matter and dark energy. Indeed, scientific theories become more ethereal and doubtful on both the smallest and largest scales, respectively nuclear physics and astronomy. An example of this, perhaps, is the unlikely hypothesis that dark matter is matter that is invisible to us because it is speeding away from us faster than the speed of light (Dannon 2013).

On the smallest scale we must content ourselves with the Bohr-Rutherford model of the atom and the sub-particles now considered to make up the basic particles of this. Nevertheless, this model works well in explaining chemical and nuclear reactions. On the larger scale, however, we have only obtained a basic view of the universe so far as we can observe, and our understanding of the universe is still relatively primitive and full of errors, and thus in need of many corrections.

Some examples of the many unanswered questions and corrections needed are:

➢ The origin of life on Earth: the possibility that this came from outer space aboard meteors cannot be discounted and requires further consideration.

Appendix E

- Human evolution: are we now in reverse evolution and thus becoming worse both physically and mentally? (Mohr 2012).
- Thanks to pollution, desertification and exploitation many animal species are likely to become extinct in the next few hundred years. Thanks to these factors and our excessive population mankind may also become extinct within a millennium, or at least our population is likely to be greatly reduced (Mohr 2012).
- The sun has a limited lifetime, after which life on Earth will cease. Will life remain on other planets or develop subsequently?
- The special theory of relativity makes absurd predictions such as clocks moving at two different velocities both running slower compared to the other, whereas any sensible theory must guarantee simultaneity of events.
- The general theory of relativity assumes a curved space-time because of the effects of gravity. This 'bent' view of things is unnecessary, as illustrated in Appendices B and C.
- Hubble's law holds that the further away a distant galaxy, the greater its recession velocity. This assumes that redshift is caused by the Doppler effect but is an absurd result that must be reviewed.
- As discussed in Chapter 9, redshift is most likely caused by cosmic dust, and gravitation may contribute also.
- As discussed in Chapter 11, cosmic microwave background (CMB) radiation is often considered to be an after-effect of the Big Bang, but is more likely the result of interaction of light with dust and/or hydrogen atoms in the interstellar medium.
- The steady-state universe theory was modified to incorporate C-fields to maintain a constant density of matter in an expanding universe. If redshift is caused by the interstellar medium, as seems most likely, the universe is not expanding nearly as fast as many think. Thus both the issue of how fast the universe is expanding, if at all, and that of whether there is constancy of matter or 'matter/energy' in the universe, requires further clarification.

Many Remaining Questions

- ➢ The Big Bang theory has required many modifications, particularly inclusion of a short period of inflation. Along with theories of creation by God or aliens, and cyclic universe theories, the Big Bang theory is now too far fetched and a more sensible theory based on what we can infer by observation, and not on guesses beyond this, is needed. The 'evolving universe' theory presented in this book is a modest attempt at laying the foundation for such a theory
- ➢ A key prediction of the evolving universe theory is that star formation rates that have declined for some 10+ billion years will plateau at some still significant level. More detailed data is needed to help confirm this prediction.
- ➢ Our understanding of black holes, quasars, and pulsars remains limited.
- ➢ The existence of dark matter and dark energy remains to be proven.
- ➢ The weight of evidence suggests that the universe is flat and infinite but further confirmation of this is needed.
- ➢ There is some evidence that the speed of light in a vacuum can be exceeded, and that photons have some mass under certain circumstances. These matters require further investigation.
- ➢ Establishing the speed of gravity and how it is propagated is another important issue, one which was briefly discussed in Appendix B.

Clearly much remains to be done to improve our understanding of the universe. Some idea of its future, however, is clearly more important than its past, but hopefully the eventual equilibrium or steady state of the universe predicted by the 'evolving universe' model will come to pass.

Glossary

Absolute magnitude. The apparent magnitude a star would have if at a distance of 10 parsecs.
Absolute zero. Zero temperature on the Kelvin scale, at which a substance has no heat energy.
A.C. Alternating (electric) current.
Alpha particle. Nucleus of a helium atom. Emitted as a result of radioactive decay of radium.
ALT. Alternative Lorentz transformation (Buenker 2007).
Apparent magnitude. Brightness of an object seen from Earth.
Archaea. Evolved from prokaryotes to use glycolysis to obtain energy.
Archean eon. From 4 billion until 2.5 billion years ago.
Asteroid. Small celestial bodies composed of rock and metal that move around the sun, especially the asteroid belt between the orbits of Mars and Jupiter.
Astronomical unit (AU). Mean distance between Earth and the Sun (approx. 93 million miles or 150 million kilometres).
Baryon. Proton or neutron.
Big Bang theory. Theory that the universe began with the explosion and rapid expansion of a very massive but infinitely small and infinitely dense 'particle'.
Big Crunch. Singularity in which the universe collapses.
Bion. Almost indestructible microbes found in cancer tumours thought to be halfway between inorganic and organic.
Black body. Hypothetical perfect absorber and radiator of energy, with no reflecting power.
Black hole. Collapsed star region with very high gravity.

Glossary

Boson. Third type of fundamental particle, the other being quarks and leptons. The three types of boson are: photons, gluons, and weakons (weak bosons). Photons transmit electromagnetic radiation, gluons transmit forces between quarks, and weakons change one type of particle into another.

Brane. Short for membrane in string theory.

Brown dwarf. Object with insufficient mass to burn hydrogen like a star. Some have orbiting planets.

CCC. Conformal cyclic cosmology model: the universe iterates through an infinite number of cycles.

Cepheid variable. Bright yellow variable star that pulsates regularly.

Chlorophyll. One of four types of green pigment responsible for photosynthesis in plants.

CERN. European Organization for Nuclear Research.

C-field. The matter creation field which keeps the average density of the universe constant in the steady-state theory.

CIB. Cosmic infrared background radiation.

Cosmic dust. Small dust particles found throughout space.

Cosmic ray. Radiation with high energy and penetration that reaches the earth from all directions.

Cosmic microwave background (CMB). Highly isotropic microwave radiation at 2.7 °K believed by proponents of the Big Bang theory to have originated soon after the Big Bang.

Cosmological constant. Constant in general theory of relativity representing a force to balance the attraction of gravitation whose value = 1 for a static universe.

Creationism. Literal belief in the account of creation in the book of Genesis.

Cyanobacteria. Mostly photosynthetic prokaryotic organisms which occur singly or in colonies. Also known as blue-green algae.

Cycad. Tropical plant with unbranched stems and a crown of fernlike leaves.

Dark matter. Non-luminous material in space predicted by cosmologists to account for unexplained gravitational effects in the universe.

Dark energy. Postulated to account for acceleration in the expansion of the universe.

Glossary

Degrees of freedom (d.f.). The variables used at each node in finite element method analysis. In plane stress analysis these are the translational displacements u and v parallel to the x and y axes.

Doppler effect. Increase (or decrease) in the frequency of sound, light, or other waves as the source and observer move towards (or away) from each other.

Electron. Small negatively charged particle with little mass the moves outside atomic nuclei. Now also known as the electron-neutrino. *See also* leptons.

Electromagnetic radiation. Radiation caused by acceleration of an electric charge, including visible light, radio waves, gamma rays, and X-rays. Causes electric and magnetic fields to vary simultaneously.

Emf. Electromagnetic force between charged particles.

Eukaryotes. Organisms with a membrane-bound nucleus containing chromosomes.

Event horizon. The boundary of a black hole.

Fermion. Any of several subatomic particles with half-integral spin, e.g. nucleons.

Galaxy. Collection of star systems.

Gamma ray. Very short length electromagnetic radiation.

General theory of relativity (GTR). Extension of the special theory of relativity to include the effects of gravity.

Glycolysis. Breaking down of carbohydrates and sugars to store energy in the bonds of adenosine triphosphate (ATP).

GPS. Global positioning satellite.

Graviton. Predicted type of boson that transmits gravity.

GRS. Gravitational redshift.

Gluon. *See* boson.

Gymnosperm. Plants with seeds not enclosed in an ovary. h^2 **extrapolation.** Assumes solution error proportional to the square of the step size in a numerical solution.

Hadean eon. Period dating from the formation of the Earth 4.6 billion years ago until 4 billion years ago.

Higgs boson. Gives particles their mass.

Hominidae. Modern man and his ancestors.

Homo habilis. First *Homo species* circa 2 million years BC.

Glossary

Homo ergaster. First *Homo sapiens* species. Evolved in Africa circa 1.5 million years BC. Migrated to Europe and evolved into *Homo Heidelbergensis*, which in turn evolved into Homo Neanderthalis and also the Denisovans.

Homo erectus. Evolved from Homo ergaster migration into Asia. Evolved into *Homo sapiens sapiens* (modern man).

Horizon problem. The limit of how much of the solar system we can see because of the finite speed of light.

Hubble's law. Law including Hubble's constant in which the velocity of distant objects in space is proportional to their distance from Earth.

Inflation. Short period in which the universe expanded from a size like that of a tennis ball to the size we see now.

Intelligent design. Modern version of creationism holding that our complex world was designed by a higher being.

IR. Infrared radiation

ISM. Interstellar medium.

Isotope. Atoms with additional neutrons in their nuclei. Some isotopes of heavier elements are radioactive.

K. Symbol for kelvin (absolute temperature).

Kuiper belt. Disk-shaped region of minor planets outside the orbit of Neptune which includes many asteroids.

LCC. Large curvature correction, a factor used in iterative residual load calculation of curvatures in beams etc. undergoing displacements.

Leptons. Fundamental particles of which there are six types: the electron and its two unstable forms, the muon and tau, and the electron neutrino and its two unstable forms, the mu neutrino and tau neutrino.

Light year. Distance travelled by light in a vacuum in one year = 9.46 trillion kilometres.

Lorentz-Fitzgerald contraction. Hypothetical contraction in length observed in a fast-moving object by a stationary observer.

Lorentz transformation. Factor involving the ratio of an object's speed to that of light and used in the general theory of relativity.

Magnetic monopole. A hypothetical isolated magnetic pole.

Meson. Unstable particles composed of even numbers of quarks and antiquarks. Pi mesons (pions) help bind nuclei.

Meteorite. Stony remnant part of a meteoroid.

Glossary

Metric expansion of space. Increase in distance between parts of the universe without those parts changing their locations. A key feature of Big Bang cosmology.

Micron. One millionth of a metre.

Microzyma. Seemingly indestructible basic form of life found in bacteria, amoebas and able to ferment and transform into bacteria.

Milky Way. Galaxy containing our solar system.

Molecular cloud. Dense concentration of gas and dust. Often found at the centres of dark nebula.

MW. Microwave radiation.

NASA. National Aeronautics and Space Administration.

Nebula. Large cloud of mostly hydrogen gas and cosmic dust. The three types are: reflection nebulae, emission nebulae, and dark nebulae.

Neutron. Particle in the nucleus of comparable mass to the proton, but without charge.

Neutron star. Small very dense star composed mainly of neutrons. These have very strong magnetic fields.

Neutrino. Small almost massless particle in nuclei which has no charge. Now also known as the electron-neutrino. *See also* leptons.

Nova. Very bright, short-lived variable star.

Nucleon. A constituent (proton or neutron) of an atomic nucleus.

Nucleosynthesis. Synthesis on a cosmic scale of chemical elements from the simplest atomic nuclei.

Olbers's paradox. That in a static infinite universe the night sky should be bright because light from nearer regions of space would be the similar to that from distant regions.

PAHs. *See* polycyclic aromatic hydrocarbons.

Parallax. Apparent difference in the position or direction of an object when viewed from different positions. In astronomy usually the angular amount of this difference.

Parsec. Astronomical unit = distance from Earth at which stellar parallax is 1 second of arc = 3.262 light years.

Periodic table. Table listing the known chemical elements, beginning with hydrogen and ending with unstable elements.

Photon. *See* boson.

Pion. *See* meson.

Glossary

Planck's constant. The constant of proportionality relating the energy of a photon to its frequency and thence its wavelength.

Plasma. Gas becomes a plasma when it is heated until the atoms lose their electrons. Applies to interstellar gas.

Pleomorphic. Able to take different forms.

Polycyclic aromatic hydrocarbons (PAHs). Organic compounds with long molecular chains. An example is benzapyrene, product of combustion of organic substances. Like several other PAHs, it is carcinogenic.

Positron. An antielectron. Interaction of a positron and an electron results in annihilation.

Prokaryotes. Unicellular organisms lacking membrane-bound nuclei. Usually bacterial but also included are blue-green algae and actinomycetes and mycoplasma.

Proton. Positively charged particle in the nucleus of atoms.

Pulsar. Rapidly spinning neutron star that emits powerful, pulsating radio waves.

QED (quantum electrodynamic) theory. Quantum theory of capture and re-emission of redshifted photons by cosmic dust.

QSSC. Quasi-steady state cosmology.

Quantum. Indivisible unit in which electromagnetic radiation is emitted or absorbed.

Quantum mechanics. Theory developed from Planck's quantum principle and Heisenberg's uncertainty principle (indeterminacy of both position and velocity of a particle).

Quarks. Fundamental particles of which there are six types: up, down, charm, strange, top, bottom. Only 'up' and 'down' quarks are stable.

Quasar. Large starlike object that emits radio waves and may send out other forms of energy with large redshifts.

Reciprocity (principle of). In the theoretical mechanics of structures this allows interchange of different incremental force and displacement terms on the basis that the same overall energy of the system is conserved.

Red giant. Very large but relatively cool star.

Redshift. Shift in spectra (= Z) of very distant galaxies toward longer red wavelengths which is generally interpreted as evidence that the universe is expanding.

Glossary

Singularity. Point in space-time at which the curvature of four-dimensional space-time becomes infinite.

Solar mass. The mass of the sun.

Special theory of relativity (STR). Theory predicting changes in length, mass, and time in a fast-moving object relative to a stationary frame of reference.

Spin. Property of subatomic particles with few discrete values.

Sporangia. Organ containing or producing spores.

Steady-state theory (of the universe). Theory that the universe remains the same on a macroscopic scale, the original theory being modified to allow for expansion.

Stomata. Epidermal pore in a leaf or stem through which gases and water vapour can pass.

Sunyaev-Zel'dovich effect (SZE). Scattering of background radiation in wavelength by collisions with electrons in large clouds of intergalactic gas.

Supernova. A star that explodes and becomes extremely luminous in the process.

String Theory. Postulates that subatomic particles are one-dimensional strings.

Strong force. The strongest of the four fundamental forces with the shortest range that binds quarks, protons and neutrons to form nuclei.

Supergiant. Very large, short-lived star with high luminosity.

Supernova. Large, aging star that explodes, becoming highly luminous.

Thermonuclear fusion. Reactions involving combination of subatomic particles, whereas nuclear fission involves splitting of nuclei.

Time dilation. Slowing of time in a fast-moving frame of reference predicted by the Special Theory of Relativity.

Tired-light effect. Weak gravitational interactions of photons with stars and other material that reduces their momentum, resulting in redshift.

Tolman Test. Surface brightness test that compares the surface brightness of galaxies as a function of redshift Z.

Tubercle. The lesion or swelling caused by tuberculosis.

UV. Ultraviolet radiation.

Glossary

White dwarf. Small very dense stars.

X-ray. Short wavelength electromagnetic radiation produced by impact of high-speed electrons.

References

Anthony, S. 'Astrophysicists create the first accurate map of the universe: It's very flat, and probably infinite'. Article posted at www.Extreme.com on Jan. 9, 2014.

Assis A. K. T., Reeves M. C. D., 'History of the 2.7 K Temperature Prior to Penzias and Wilson', www.redshift.vif.com/JournalFiles/Pre2001/V02NO3/V02N3ASS.PDF.

BBC, article at www.bbc.co.uk/nature/history_of_the_earth, viewed Feb. 2014.

Bell E. F., Papovich C., Wolf C., Le Floc'h E., Caldwell J. A. R., Barden M., Egami E., McIntosh D. H., Meisenheimer K., Perez-Gonzalez P. G., Rieke G. H., Rieke M. J., Rigby J. R., Rix H. W., 'Toward an understanding of the rapid decline of the cosmic star formation rate', *The Astrophysical Journal,* 625/1 (2005).

Bernstein J. *Einstein.* Fontana, London (1973).

Biba, E., *Space travel means that the threat of foreign germs coming to Earth is all too real,* article posted on a NASA website, July 25, 2012.

Bruce Brown, Lane Morgan, *The Miracle Planet,* Child and Associates, Frenchs Forest NSW (1989).

Buenker, R. J., *Simultaneity and Relativity: The Alternative Lorentz Transformation,* simultaneity-relativity.blogspot.com.au/(June 2007).

Buenker, R. J., *Logical Basis of Fitzgerald-Lorentz Contraction, Wikipedia* page posted April 24, 2011 at en.wikipedia.org/wiki/User:Rjbeunker.

Buenker, R. J., 'The myth of Fitzgerald-Lorentz length contraction and the reality of Einstein's velocity transformation', *Aperion,* vol. 20, no. 1 (2013) pp 27-52.

References

Burbidge E. M., Burbidge G. R., Fowler W. A., Hoyle F., Synthesis of the Elements in Stars, *Reviews in Modern Physics* no. 29 (1957) pp 547-650.

Burchell, Bernard, *Experimental Evidence for Time Dilation,* www.alternativephysics.org/book/TimeDilationExperiments.htm.

Cantwell A., *The Cancer Microbe.* Aries Rising Press, LA (1990).

Chaline, E, *History's Worst Predictions, and the People Who Made Them.* The History Press, Briscombe Port, Stroud (2011).

Cullen, H, *The Weather of the Future, Heat Waves, Extreme Storms, and Other Scenes From a Climate-changed Planet.* Harper, New York (2010).

Dannon H. V., 'A hypothesis that dark matter is regular matter, invisible, since it moves faster than light'. *Gauge Institute Journal* 9 (2013) 4.

Ebor D., *The New English Bible,* Oxford University Press, Oxford (1972).

Eddington A. S., *The Mathematical Theory of Relativity,* Cambridge University Press, Cambridge, 2nd ed. (1924).

Egerton Eastwick R. W. (ed.), *The Oracle Encyclopaedia,* George Newnes, London (1896).

Einstein, A., *The Meaning of Relativity.* Methuen, London (1922).

Frank, A. *3 Theories That Might Blow Up the Big Bang,* article posted at discovermagazine.com/ . . ./25-3-theories-that-might-blow-up-the-big-bang in March 2008.

Frim Physics, *special relativity disproved,* www.math.ucr.edu/home/baez/physics/Relativity/SpeedofLight/speed_of _light.html (May 2013).

Frisch P. C., Bzowski M., Livadiotis G., McComas D. J., Moebius E., Mueller H. R., Pryor W. R., Schwadron J. M., Vallerga J. V., Ajello J. M., 'Decades-long changes of the interstellar wind through our solar system', *Science* 341 (2013) 6150.

Gaines M., *Atomic Energy,* Grosset & Dunlap, New York (1970).

Gribbin, J., *Q is for QUANTUM: Particle Physics from A to Z,* Weidenfeld & Nicolson, London (1998).

Hartnett, J., 'The Big Bang fails another test', article posted on the internet at creation.com/the-big-bang-fails-another-test, 15 September 2006, citing "NASA/WMAP Science Team".

Hartnett, J., 'Evidence against expansion', article posted on creation.com/expanding-universe.2, viewed Jan. 2014.

References

Hawking S., *A Brief History of Time: From the big bang to black holes,* Bantam/Transworld, London (1989).

Hoyle F., *The Intelligent Universe: A New View of Creation and Evolution,* Michael Joseph, London (1983).

Hoyle and Wickramasinghe, 'The ultraviolet absorbance of presumably interstellar bacteria and other matters', *Astrophysics and Space Science* vol. 111, issue 1, pp 65-78 (1985) available on the Internet via Springer Link.

Hoyle F., Burbidge G., Narlikar J. V., 'A quasi-steady state cosmological model with creation of matter', *The Astrophysical Journal,* vol. 410 (1993) pp 437-457.

Humphreys, D. R., *Bumps in the Big Bang,* article posted on the internet at www.icr.org/article/bumps-big-bang.

Jayawardhana R., *Strange New Worlds: The search for alien planets and life beyond our solar system,* Princeton University Press (2011).

Jones, B., *The Beginner's Guide to Astronomy,* Artist's House/Mitchell Beazley Int., London (1987).

Jones L. V., *Stars and Galaxies,* Greenwood Press/ABC-CLIO, Santa Barbara CA (2010).

Kistler M. D., Yuksel H., Hopkins A. M., The cosmic star formation rate from the faintest galaxies in the unobservable universe, UC Berkeley report no. UCB-NPAT-13-003, Internet ref. arXiv:1305.1630v.

Klyce, B., *Bacteria: The space colonists,* article on the Cosmic Ancestry website at www.panspermia.org/bacteria.htm

Kramer, M., article re biofilms on the space shuttle Atlantis, www.space.com/21886-space-bacteria-grows-strange-ways

Kreyszig, E., *Advanced Engineering Mathematics,* 4th ed., Wiley, New York (1979).

Lerner E., 'Bucking the big bang', *New Scientist* 182 (May 2004) 20.

Leuchs G, Sanchez-Soto L. L., 'A sum rule for charged elementary particles', *European Physical Journal D* (2013), DOI: 10.1140/epjd/e2013-30577-8 (open access).

Lofts G., Preuss P., Gilbert K., *Living With Science and Technology* (book 2), The Jacaranda Press, Milton QLD (1991).

Madeiros R. W., *Chemistry, An Interdisciplinary Approach,* Van Nostrand Reinhold, New York (1971).

References

May B., Moore P., Linkott C., *Bang! The Complete History of the Universe,* Cameron House, Wingfield SA (2006).

Millenium Group, 'Transformations of the solar system', www.tmgnow.com/repository/global/planetophysical1.html.

Mohr G. A., Milner H. R., 'Finite element analysis of large displacements in flexural systems', *Computers and Structures,* vol. 13 (1981) pp 533-536.

Milner H. R., 'Accurate finite element analysis of large displacements in skeletal frames', *Computers and Structures* vol. 14 (1981) pp 205-210.

Mohr G. A., Argyris J. H., 'The large curvature correction in finite element analysis II', *International Journal of Arts and Sciences,* vol. 1, no. 2 (2001) pp 27-35.

Mohr G. A., *The Doomsday Calculation,* Xlibris, Sydney (2012).

Mohr G. A., *Heart Disease, Cancer, and Aging: Proven Neutraceutical and Lifestyle Solutions,* Horizon Publishing Group, Sydney (2013a).

Mohr G. A., *The Pretentious Persuaders* 2nd ed, Horizon Publishing Group, Sydney (2013b).

Mohr G. A., Fear Edwin, *World Religions: The history, psychology, issues, and truth,* Horizon Publishing Group, Sydney (2014).

Muller F. J., 'An experimental disproof of special relativity theory', article at worldnpa.org/pdf/abstracts/abstracts-113.pdf.

Muir, H., 'Does the universe go on forever?' *New Scientist,* 11th October 2003 (www.newscientist.com).

NASA, 'Will the Universe expand forever?', www.map.gsfc.nasa.gov/universe/uni-shape.html.

Noel, D., *R.I.P. expanding universe (b. 1930, d. 2012): The Big Bang Never Happened,* article posted on the internet in 2012 at aoi.com.au/bcw/RIPExpanding.

Noel, D., 'A new model for the origin of cosmic microwave background radiation', internet article at aoi.com.au/scientific/SciCMBR.

Noel, D., 'The placid universe model', internet article at www.aoi.com.au/bcw/Placid.

Oerter R., *The Theory of Almost Everything,* Pi Press, New York (2006).

References

Palumbi S. R., *The Evolution Explosion, How Humans Cause Rapid Evolutionary Change,* WW Norton, New York (2001).

Penrose, R., *Cycles of time: An extraordinary view of the universe,* The Bodley Head, London (2010).

Prevenslik, T., 'Redshift by cosmic dust trumps Hubble and tired light theories', www.news-about-space.org/astronomy-news/cluster5307942

Prokhovnik, S. J., *The Logic of Special Relativity,* Melbourne University Press, Melbourne (1967).

Renshaw, C., 'The direct verification of length contraction and time dilation in modern satellite systems and cosmological studies', article posted on the internet and at www.renshaw.teleinc.com/papers/lonon1/london.stm.

Ricker, H. R., 'Einstein's false derivation of Lorentz-Fitzgerald contraction', *The General Science Journal,* paper posted on the Internet at www.gsjournal.net/old/science/ricker13.pdf

Ridpath, I. (ed.), *The Illustrated Encyclopedia of the Universe,* Watson-Guptill, New York (2011).

Shaunacy, 'A rare new bacterium thrives in spacecraft cleanrooms', article posted on the Internet at www.popsci.com/.../rare-new-bacterium-thrives-spacecraft-cleanrooms.

Singh S., *Big Bang (The most important scientific discovery of all time and why you need to know about it),* Harper Perennial, London (2005).

Smid, T., 'Algebraic Time Dilation Paradox', article posted at Physicsmyths.org.uk/timedilation.htm.

Smith C. M., Davies E. T., *Anthropology for Dummies,* Wiley, New Jersey (2008).

Smithsonian Institute, 'The First Life on Earth', paeleobiology.si.edu/geotime/main/htmlversion/archean3.html.

Space.com, *Star formation sputtering out across the universe,* space.com/18370-universe-star-formation-rate-decline.html.

Sparrow, G., *Probing Deep Space,* Hinkler Books, Heatherton VIC (2006).

Sparrow, G., *Cosmos Close-Up,* Quercus Books (2013), www.quercusbooks.co.uk.

Sungenis R., 'A disproof of the special theory of relativity', galileowaswrong.com/.../A-Disproof-of-the-Special-Theory-of-Relativity.

References

Sykes, N., 'A new paradigm of thought: The electric universe (a view from the Cayman Islands)', www.churchofenglandcayman.com/SciCom11.html.

Thomas, B., 'Viral life from outer space? Not likely', article posted on the Institute for Creation Research website, viewed Feb. 2014.

Urban M. et al., 'The quantum vacuum as the origin of the speed of light', *European Physical Journal D* (2013), DOI: 10.1140/epjd/e2013-30578-7 (open access). van Flandern, T., 'The speed of gravity: What the experiments say', *Physics Letters A,* 250 (1998),pp1-11 (article also posted on the Internet). van Lawick-Goodall, J., *In The Shadow of Man*, Houghton Mifflin, Boston (1971).

Weiss M. L., Mann A. E., *Human Biology and Behaviour, An Anthropological Perspective*, 2nd ed, Little Brown, Boston MA (1978).

Weinberg R., *One Renegade Cell*, Phoenix, London (1999).

Wetterich, C., paper in the journal *Nature*, doi:10.1038/nature.2013.13379, preprint available at www.arxiv/org/abs/1303.6878 (2013).

General References

Chambers Dictionary of World History, Chambers Harrap, Edinburgh (1993).

Collins National Encyclopedia, Collins, London (1966).

Encarta Book of Quotations, Pan MacMillan, Sydney (2000).

Encarta Encyclopedia 1999, Microsoft Corporation, 1998.

Encyclopedia Britannica CD 99 (1999).

Encyclopedia Britannica Ready Reference, 2003.

The 1996 Grolier Multimedia Encyclopedia, Grolier Inc., Grolier Electronic Publishing Inc., New York (1996).

Microsoft Bookshelf '94.

Mindscape World Atlas & Almanac, The Learning Company Inc., Crawley, West Sussex (1999).

Natkeil R., Atlas of 20th Century History, Bison Books, Greenwich CT (1982).

The Oxford Dictionary of Quotations. 4th ed., Oxford University Press, Oxford (1992).

References

The Oxford Interactive Encyclopedia, The Learning Company (1997).
Concise Oxford Dictionary, 9th ed. (CD edition), Oxford University Press, Oxford (1999).
Philips' New World Atlas, George Philip & Son, Ltd, London (1934).
Power Quotes, Daniel B. Baker, The Business Library, Information Australia (1992).
The SBS World Guide, 4th ed., Reed Reference Australia, Melbourne (1995).
The World Almanac and Book of Facts 1998, World Almanac Books, Mahwah NJ (1998),
Wikipedia—various articles, mostly during 2013 & 2014.
World Book 1999—Multimedia Encyclopedia, World Book Inc. (1998).
Wordweb 6, copyright Anthony Lewis (2009), database copyright Princeton University (2006).

www.ingramcontent.com/pod-product-compliance
Lightning Source LLC
Chambersburg PA
CBHW030935180526
45163CB00002B/569